Study Guide

Donna Friedman

St. Louis Community College
Florissant Valley

Introductory Chemistry

SECOND EDITION

Nivaldo J. Tro

PEARSON
Prentice Hall

Upper Saddle River, NJ 07458

Project Manager: Kristen Kaiser
Executive Editor: Kent Porter-Hamann
Editor-in-Chief, Science: John Challice
Executive Managing Editor: Kathleen Schiaparelli
Assistant Managing Editor: Becca Richter
Production Editor: Dana Dunn
Supplement Cover Manager: Paul Gourhan
Supplement Cover Designer: Joanne Alexandris
Manufacturing Buyer: Alan Fischer
Cover Image: Quade Paul

PEARSON
Prentice Hall

© 2006 Pearson Education, Inc.
Pearson Prentice Hall
Pearson Education, Inc.
Upper Saddle River, NJ 07458

All rights reserved. No part of this book may be reproduced in any form or by any means, without permission in writing from the publisher.

Pearson Prentice Hall™ is a trademark of Pearson Education, Inc.

The author and publisher of this book have used their best efforts in preparing this book. These efforts include the development, research, and testing of the theories and programs to determine their effectiveness. The author and publisher make no warranty of any kind, expressed or implied, with regard to these programs or the documentation contained in this book. The author and publisher shall not be liable in any event for incidental or consequential damages in connection with, or arising out of, the furnishing, performance, or use of these programs.

> **This work is protected by United States copyright laws and is provided solely for teaching courses and assessing student learning. Dissemination or sale of any part of this work (including on the World Wide Web) will destroy the integrity of the work and is not permitted. The work and materials from it should never be made available except by instructors using the accompanying text in their classes. All recipients of this work are expected to abide by these restrictions and to honor the intended pedagogical purposes and the needs of other instructors who rely on these materials.**

Printed in the United States of America

10 9 8 7 6 5 4 3 2 1

ISBN 0-13-147071-X

Pearson Education Ltd., *London*
Pearson Education Australia Pty. Ltd., *Sydney*
Pearson Education Singapore, Pte. Ltd.
Pearson Education North Asia Ltd., *Hong Kong*
Pearson Education Canada, Inc., *Toronto*
Pearson Educación de Mexico, S.A. de C.V.
Pearson Education—Japan, *Tokyo*
Pearson Education Malaysia, Pte. Ltd.

Contents

1	The Chemical World	1
2	Measurement and Problem Solving	7
3	Matter and Energy	27
4	Atoms and Elements	43
5	Molecules and Compounds	55
6	Chemical Composition	67
7	Chemical Reactions	85
8	Quantities in Chemical Reactions	105
9	Electrons in Atoms and the Periodic Table	119
10	Chemical Bonding	133
11	Gases	147
12	Liquids, Solids, and Intermolecular Forces	169
13	Solutions	185
14	Acids and Bases	203
15	Chemical Equilibrium	219
16	Oxidation and Reduction	235
17	Radioactivity and Nuclear Chemistry	249
18	Organic Chemistry	263
19	Biochemistry	281

The Chemical World

1

CHAPTER OVERVIEW

Chapter 1 introduces chemistry using familiar examples. Ordinary substances are linked to the atoms and molecules that compose them. The scientific method is described.

CHAPTER OBJECTIVES

After reading and studying the text, students should be able to:

1. Give examples of ordinary things composed of chemicals.
2. Define chemistry.
3. List the steps in the scientific method.
4. Differentiate between observations, laws, theories, and experiments.
5. State the law of conservation of mass.

CHAPTER IN REVIEW

- Virtually everything is composed of chemicals. Chemicals are all around us.

- Chemistry is the study of atoms and molecules. By studying atoms and molecules, chemists learn why matter behaves as it does.

- The scientific method consists of observations, laws, theories, and experiments.

- Observations involve noting or measuring some aspect of nature.

- Laws summarize the results of a large number of observations.

- Theories are models that give the underlying causes for observations and laws.

- Experiments test theories.

- If an experimental result is inconsistent with a theory, the theory must be revised.

Chapter 1

SELF-TEST QUESTIONS

A. Match the following terms with the phrases below.

  atomic theory   scientific law
  law of conservation of mass scientific method

1. Process of learning that emphasizes observation and experimentation
2. Matter is neither created nor destroyed in a chemical reaction
3. All matter is composed of small indestructible particles called atoms
4. General statement that summarizes past observations and predicts future ones

B. True/False

1. Water is a chemical.
2. Energy has mass.
3. All chemicals are poisonous.
4. Molecules are composed of atoms.
5. Chemical bonds hold atoms together.
6. Bubbles in soda pop are pockets of carbon dioxide gas.
7. Water molecules are composed of one oxygen atom and one hydrogen atom.
8. Antoine Lavoisier formulated atomic theory.
9. The scientific method begins with observation.
10. Experiments are designed to prove theories.

C. Multiple Choice

1. Chemicals are present in
    a) The air we breathe
    b) The water we drink
    c) The food we eat
    d) All of the above

2. When a can of soda pop is opened
    a) The pressure inside the can increases
    b) Carbon dioxide molecules escape
    c) Sugar molecules escape
    d) Water bubbles form

3. To understand the characteristics of octane, a component of gasoline, chemists investigate the properties of
    a) Methane molecules
    b) Water molecules
    c) Ethanol molecules
    d) Octane molecules

4. Carbon dioxide molecules are composed of
   a) Carbon dioxide atoms
   b) Carbon molecules and oxygen molecules
   c) Carbon atoms and oxygen molecules
   d) Carbon atoms and oxygen atoms

5. "A yellow solid settled at the bottom of the liquid" is an example of a(n)
   a) Observation
   b) Experiment
   c) Law
   d) Theory

6. "Gases are composed of widely spaced noninteracting particles" is an example of a(n)
   a) Observation
   b) Law
   c) Theory
   d) Measurement

7. Experiments are designed to
   a) Prove theories
   b) Test theories
   c) Discard theories
   d) Violate laws

8. The scientific method does not apply to which discipline?
   a) Physics
   a) Chemistry
   b) Astronomy
   d) Philosophy

9. Combustion refers to
   a) Melting
   b) Boiling
   c) Burning
   d) Breaking apart

10. Which of the following is a qualitative observation?
    a) Aluminum melts at 658 °C.
    b) Aluminum has a specific heat capacity of 0.903 J/g °C.
    c) Aluminum conducts electricity.
    d) Aluminum has a density of 2.7 g/cm^3.

Chapter 1

D. Crossword Puzzle

ACROSS

1. Science that seeks to understand what matter does by studying what atoms and molecules do
3. Description of underlying reasons for observations and laws
4. Measurement of an aspect of the physical or natural world
5. Well-established model that predicts behavior well beyond the observations and laws from which it is formed

DOWN

2. Test that looks for observable predictions of a theory

ANSWERS TO SELF-TEST QUESTIONS

A. Matching
1. scientific method 2. law of conservation of mass 3. atomic theory 4. scientific law

B. True/False
1. T 2. F 3. F 4. T 5. T 6. T 7. F 8. F 9. T 10. F

The Chemical World

C. Multiple Choice
1. d   2. b   3. d   4. d   5. a   6. c   7. b   8. d   9. c   10. c

D. Crossword Puzzle

¹C	H	²E	M	I	S	T	R	Y			
		X									
³H	Y	P	O	T	H	E	S	I	S		
		E									
		R		⁴O	B	S	E	R	V	A	T	I	O	N
		I									
		M									
⁵T	H	E	O	R	Y						
		N									
		T									

5

Chapter 1

Measurement and Problem Solving   2

CHAPTER OVERVIEW

Chapter 2 focuses on numbers and measurements. Scientific notation is discussed. Uncertainty in measurements and conventions for reporting measured values are examined. Use of significant figures in calculations is explained. The International System of units is introduced. Unit conversions are emphasized. Numerical problem-solving strategies are presented.

CHAPTER OBJECTIVES

After reading and studying the text, students should be able to:

1. Express numbers in scientific notation.
2. Convert numbers written in scientific notation to decimal notation.
3. Determine the number of significant figures in a number.
4. Differentiate between measured numbers and exact numbers.
5. Report measured quantities to the correct number of significant digits.
6. Round calculated numbers to the correct number of significant digits.
7. List the SI standard units of length, mass, time, and temperature.
8. List the SI prefix multipliers, their symbols, and numerical meanings.
9. Convert from one unit to another.
10. Solve multistep conversion problems.
11. Convert quantities involving units raised to a power.
12. Define density.
13. Calculate the density of a substance given its mass and volume.
14. Use density to convert from mass to volume or from volume to mass.

CHAPTER IN REVIEW

- A unit is a standard quantity by which other quantities are measured.

- Scientific notation is used to write very small and very large numbers. It consists of a decimal part and an exponential part.

- To write a number in scientific notation, move the decimal point to obtain a number between 1 and 10, multiply the result by a power of 10. The exponent is positive if the decimal point was moved to the left and negative if the decimal point was moved to the

Chapter 2

right. The absolute value of the exponent is the number of places the decimal point was moved.

- Scientific numbers are reported so that every digit is certain except the last, which is estimated.

- Non-place-holding digits in a reported measurement are significant digits.

- To determine the number of significant digits in a number, count the number of digits starting with the left-most nonzero digit. Zeros at the end of a number that does not have a decimal point are ambiguous; such numbers should be written in scientific notation.

- Exact numbers have an unlimited number of significant digits. Exact numbers are counted numbers and numbers that originate from definition.

- In multiplication and division, the result should carry the same number of significant figures as the factor with the least number of significant digits.

- In addition or subtraction, the result should carry the same number of decimal places as the quantity with the fewest decimal places.

- In calculations involving both multiplication/division and addition/subtraction, do the steps in parentheses first. Determine the number of significant digits in the intermediate answer, then do the remaining steps.

- When rounding to the correct number of significant digits, round down if the last (or left-most) digit being dropped is four or less, round up if the last (or left-most) digit being dropped is five or more.

- The International System of units is a set of standard units agreed upon by scientists throughout the world. SI units include the meter (m), kilogram (kg), second (s), and kelvin (K).

- The International System of units uses prefix multipliers with the standard units to change the values of the units by powers of ten.

- A derived unit is formed from other units. A unit of length cubed is a unit of volume.

- To solve unit conversion problems, write down the given quantity and its units. Write down the *find* quantity and its units. Write down the appropriate conversion factors. Draw a solution map. Multiply the given quantity and its units by the appropriate conversion factor(s). Give the answer with units rounded to the correct number of significant figures. Check the answer.

- The density of a substance is the ratio of its mass to its volume. Units of density are those of mass divided by those of volume.

Measurements and Problem Solving

- To calculate the density of a substance, divide the mass of any sample of the substance by the volume of the sample.

- Density can be used directly as a conversion factor to convert from volume to mass. The inverse of density is used to convert from mass to volume.

SKILLBUILDER PROBLEMS AND SOLUTIONS

SKILLBUILDER 2.1   **Scientific Notation**

The total federal debt in the year 2004 was approximately $7,132,000,000,000. Express this number in scientific notation.

Solution:
To obtain a number between 1 and 10, move the decimal point to the left 12 places; therefore the exponent is 12. Since we moved the decimal point to the left, the sign of the exponent is positive. So we write

$$\$7,132,000,000,000 = \$7.132 \times 10^{12}$$

SKILLBUILDER 2.2   **Scientific Notation**

Express the number 0.000038 in scientific notation.

Solution:
To obtain a number between 1 and 10, we move the decimal point to the right 5 places. Since we moved the decimal point to the right, the sign of the exponent is negative; therefore, the exponent is –5. So we write

$$0.000038 = 3.8 \times 10^{-5}$$

SKILLBUILDER 2.3   **Reporting the Right Number of Digits**

A thermometer is used to measure the temperature of a backyard hot tub, and the reading is shown in Figure 2.4 in the text. Write the temperature reading to the right number of digits.

Solution:
Since the mercury is between 103 °F and 104 °F, we mentally divide the space between markings into ten equal spaces and estimate the next digit. In this case, the result should be reported as

103.4 °F

Chapter 2

In general, one unit of difference in the last digit is acceptable because the last digit is estimated and different people might estimate it slightly differently.

SKILLBUILDER 2.4   **Determining the Number of Significant Figures in a Number**

How many significant figures are in each of the following numbers?
a) 58.31
b) 0.00250
c) 2.7×10^3
d) 1 cm = 0.01 m
e) 0.500
f) 2100

Solution:
a) 58.31
   All digits are significant for a total of 4 significant figures.
b) 0.00250
   The leading zeros only mark the decimal place—they are not significant. The trailing zero is significant as are the 2 and the 5 for a total of 3 significant figures.
c) 2.7×10^3
   Both digits in the decimal part are significant for a total of 2 significant figures.
d) 1 cm = 0.01 m
   Defined numbers have an unlimited number of significant figures.
e) 0.500
   The trailing zeros are significant as is the 5 for a total of 3 significant figures.
f) 2100
   The number is ambiguous. Write 2.1×10^3 to indicate 2 significant digits or as 2.100×10^3 to indicate 4 significant figures.

SKILLBUILDER 2.5   **Significant Figures in Multiplication and Division**

Perform the following calculations to the correct number of significant figures.
a) $1.10 \times 0.512 \times 1.301 \times 0.005 \div 3.4$
b) $4.562 \times 3.99870 \div 89.5$

Solution:
a) $1.10 \times 0.512 \times 1.301 \times 0.005 \div 3.4 = 0.001$ or 1×10^{-3}

   We round the intermediate answer of 0.0010775 to 1 significant figure to reflect the 1 significant figure in the least precisely known quantity (0.005).

b) $4.562 \times 3.99870 \div 89.5 = 0.204$

Measurements and Problem Solving

We round the intermediate answer of 0.203822 to 3 significant figures to reflect the 3 significant figures in the least precisely known quantity (89.5).

SKILLBUILDER 2.6   **Significant Figures in Addition and Subtraction**

Perform the following calculations to the correct number of significant figures.

a)
    2.18
 + 5.621
 + 1.5870
 −1.8

b)
    7.876
 − 0.56
 +123.792

Solution:

a)
    2.18
 + 5.621
 + 1.5870
 −1.8
    7.6

We round the intermediate answer of 7.588 to one decimal place to reflect the quantity with the fewest decimal places (1.8).

b)
    7.876
 − 0.56
 +123.792
  131.11

We round the intermediate answer of 131.108 to two decimal places to reflect the quantity with the fewest decimal places (0.56).

SKILLBUILDER 2.7   **Significant Figures in Calculations Involving Both Multiplication/Division and Addition/Subtraction**

a) $3.897 \times (782.3 - 451.88)$
b) $(4.58 \div 1.239) - 0.578$

Solution:
a) $3.897 \times (782.3 - 451.88)$

We do the step in parentheses first.

11

Chapter 2

$$= 3.897 \times (330.42)$$

We use the subtraction rule to mark 330.42 to one decimal place since 782.3, the number with the least number of decimal places, has one.

$$= 3.897 \times 330.4\underline{2}$$

We then perform the multiplication and round the intermediate answer of 1287.6467 to four significant digits since both numbers have four significant digits.

$$3.897 \times (330.\underline{4}2) = 1288$$

b) $(4.58 \div 1.239) - 0.578$

We do the step in parentheses first.

$$= (3.696529) - 0.578$$

The number with the least number of significant digits (4.58) has three, so we mark the answer to three significant figures.

$$= 3.6\underline{9}6529 - 0.578$$

We then perform the subtraction and round the answer to two decimal places since both numbers have two decimal places.

$$3.6\underline{9}6529 - 0.578 = 3.12$$

SKILLBUILDER 2.8   Unit Conversion

Convert 56.0 cm to inches.

Given: 56.0 cm

Find: in

Conversion Factor:

1 in. = 2.54 cm

Measurements and Problem Solving

Solution Map:

cm → in.

$$\frac{1 \text{ in.}}{2.54 \text{ cm}}$$

Solution:
The conversion factor is written so that cm, the unit we are converting from, is on the bottom and in, the unit we are converting to, is on the top.

$$56.0 \text{ cm} \times \frac{1 \text{ in.}}{2.54 \text{ cm}} = 22.\underline{0}4724 \text{ in.} = 22.0 \text{ in.}$$

We round to three significant figures since the quantity given has three significant figures. The conversion factor is exact.

SKILLBUILDER 2.9   Unit Conversion

Convert 5678 m to kilometers.

Given: 5678 m

Find: km

Conversion Factor:

$$1 \text{ km} = 10^3 \text{ m}$$

Solution Map:

m → km

$$\frac{1 \text{ km}}{1000 \text{ m}}$$

Solution:
The conversion factor is written so that m, the unit we are converting from, is on the bottom and km, the unit we are converting to, is on the top.

$$5678 \text{ m} \times \frac{1 \text{ km}}{1000 \text{ m}} = 5.678 \text{ m}$$

Chapter 2

We leave the answer with four significant figures since the quantity given has four significant figures. The conversion factor is a definition, which therefore does not affect the number of significant figures in the answer.

SKILLBUILDER 2.10   Solving Multistep Unit Conversion Problems

Suppose a recipe calls for 1.2 cups of oil. How many liters is this?

Given: 1.2 cups

Find: L

Conversion Factors:

    4 cups = 1 qt

    1.057 qt = 1 L

Solution Map:

$$\boxed{\text{cups}} \rightarrow \boxed{\text{qt}} \rightarrow \boxed{\text{L}}$$

$$\frac{1 \text{ qt}}{4 \text{ cups}} \quad \frac{1 \text{ L}}{1.057 \text{ qt}}$$

Solution:
Follow the solution map. Start with the quantity given and its units. Multiply by the appropriate conversion factors.

$$1.2 \text{ cups} \times \frac{1 \text{ qt}}{4 \text{ cups}} \times \frac{1 \text{ L}}{1.057 \text{ qt}} = 0.2\underline{8}3822 \text{ L} = 0.28 \text{ L}$$

We round the final answer to two significant figures since the quantity given has two significant figures.

SKILLBUILDER 2.11   Solving Multistep Unit Conversion Problems

A running track measures 1056 ft per lap. To run 15.0 km, how many laps should you run?
1 mi = 5280 ft.

Given: 15.0 km

14

Measurements and Problem Solving

Find: laps

Conversion Factors:

1 lap = 1056 ft

5280 ft = 1 mi

0.6214 mi = 1 km

Solution Map:

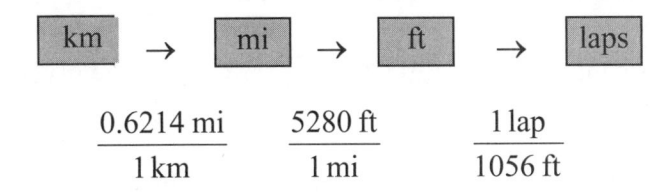

$$\boxed{\text{km}} \rightarrow \boxed{\text{mi}} \rightarrow \boxed{\text{ft}} \rightarrow \boxed{\text{laps}}$$

$$\frac{0.6214 \text{ mi}}{1 \text{ km}} \qquad \frac{5280 \text{ ft}}{1 \text{ mi}} \qquad \frac{1 \text{ lap}}{1056 \text{ ft}}$$

Solution:

$$15.0 \text{ km} \times \frac{0.6214 \text{ mi}}{1 \text{ km}} \times \frac{5280 \text{ ft}}{1 \text{ mi}} \times \frac{1 \text{ lap}}{1056 \text{ ft}} = 46.\underline{6}05 \text{ laps} = 46.6 \text{ laps}$$

We round to three significant figures since the quantity given has three significant figures.

SKILLBUILDER PLUS Solving Multistep Unit Conversion Problems

An island is 5.72 nautical mi from the coast. How far is the island in meters? (1 nautical mile = 1.151 mi)

Given: 5.72 nautical mi

Find: m

Conversion Factors:

1000 m = 1 km

0.6214 mi = 1 km

1 nautical mi = 1.151 mi

Chapter 2

Solution Map:

nautical mi → mi → km → m

$$\frac{1.151 \text{ mi}}{1 \text{ nautical mi}} \quad \frac{1 \text{ km}}{0.6214 \text{ mi}} \quad \frac{1000 \text{ m}}{1 \text{ km}}$$

Solution:
Follow the solution map to solve the problem.

$$5.72 \text{ nautical mi} \times \frac{1.151 \text{ mi}}{1 \text{ nautical mi}} \times \frac{1 \text{ km}}{0.6214 \text{ mi}} \times \frac{1000 \text{ m}}{1 \text{ km}} = 1.06 \times 10^4 \text{ m}$$

We round the answer to three significant figures to reflect the uncertainty in the given quantity.

SKILLBUILDER 2.12 **Converting Quantities Involving Units Raised to a Power**

An automobile engine has a displacement (a measure of the size of the engine) of 289.7 in^3. What is its displacement in cubic centimeters?

Given: 289.7 in.^{3}

Find: cm^3

Conversion Factor:

2.54 cm = 1 in.

Solution Map:

in.^{3} → cm^3

$$\frac{(2.54 \text{ cm})^3}{1 \text{ in.}^3}$$

Solution:
Follow the solution map to solve the problem.

$$289.7 \text{ in}^3 \times \frac{(2.54 \text{ cm})^3}{1 \text{ in.}^3} = 4747.3324 \text{ cm}^3 = 4747 \text{ cm}^3$$

We round our answer to four significant digits since the given quantity has four significant figures. The conversion factor is exact.

Measurements and Problem Solving

SKILLBUILDER 2.13   Solving Multistep Problems Involving Units Raised to a Power

How many cubic inches are there in 3.25 yd³?

Given: 3.25 yd³

Find: in.³

Conversion Factor:

36 in. = yd

Solution Map:

$$\boxed{yd^3} \rightarrow \boxed{in.^3}$$

$$\frac{(36 \text{ in.})^3}{1 \text{ yd}^3}$$

Solution:
Follow the solution map.

$$3.25 \text{ yd}^3 \times \frac{(36 \text{ in.})^3}{1 \text{ yd}^3} = 1.52 \times 10^5 \text{ in.}^3$$

We round the answer to three significant figures since the given quantity has three significant figures. The conversion factor is exact.

SKILLBUILDER 2.14   Calculating Density

A woman takes a ring back to the jewelry shop, where she is met with endless apologies. They accidentally had made the ring out of silver rather than platinum. They gave her a new ring that they promise is platinum. This time when she checks the density, she finds the mass of the ring to be 9.67 g and its volume to be 0.452 cm³. Is this ring genuine? (The density of platinum is 21.4 g/cm³.)

Given: $m = 9.67$ g

$V = 0.452$ cm³

Find: density in g/cm³

Chapter 2

If the density matches that of platinum, then the ring is probably genuine.

Equation:

$$d = \frac{m}{V}$$

Solution Map:

$$m, V \rightarrow d$$

$$d = \frac{m}{V}$$

Solution:

$$d = \frac{m}{V} = \frac{9.67 \text{ g}}{0.452 \text{ cm}^3} = 21.4 \text{ g/cm}^3$$

The density of the ring is consistent with platinum.

SKILLBUILDER 2.15   **Density as a Conversion Factor**

A drop of acetone (nail polish remover) has a mass of 35 mg and a density of 0.788 g/cm^3. What is its volume in cubic centimeters?

Given: 35 mg

Find: volume in cm^3

Conversion Factors:

0.788 g/cm^3

1 mg = 10^{-3} g

Density is a conversion factor between mass and volume. To convert from g to cm^3 we need to invert the density because we want g, the unit we are converting from, to be on the bottom and cm^3, the unit we are converting to, on the top.

Measurements and Problem Solving

Solution Map:

$$mg \rightarrow g \rightarrow cm^3$$

$$\frac{0.001 \text{ g}}{1 \text{ mg}} \quad \frac{1 \text{ cm}^3}{0.788 \text{ g}}$$

Solution:

$$35 \text{ mg} \times \frac{0.001 \text{ g}}{1 \text{ mg}} \times \frac{1 \text{ cm}^3}{0.788 \text{ g}} = 4.4 \times 10^{-2} \text{ cm}^3$$

SKILLBUILDER PLUS

A steel cylinder has a volume of 246 cm^3 and a density of 7.93 g/cm^3. What is its mass in kilograms?

Given: 246 cm^3

Find: mass in kg

Conversion Factors:

7.93 g/cm^3

1 kg = 1000 g

Solution Map:

$$cm^3 \rightarrow g \rightarrow kg$$

$$\frac{7.93 \text{ g}}{1 \text{ cm}^3} \quad \frac{1 \text{ kg}}{1000 \text{ g}}$$

Solution:
Follow the solution map to solve the problem.

$$246 \text{ cm}^3 \times \frac{7.93 \text{ g}}{1 \text{ cm}^3} \times \frac{1 \text{ kg}}{1000 \text{ g}} = 1.95 \text{ kg}$$

Chapter 2

SKILLBUILDER 2.16 **Unit Conversion**

A pure gold metal bar displaces 0.82 L of water. What is its mass in kilograms? (The density of gold is 19.3 g/cm^3.)

Given: 0.82 L water

Find: kg gold

Conversion Factors:

19.3 g/cm^3

1 mL = 10^{-3} L

1 mL = 1 cm^3

1 kg = 1000 g

Solution Map:

$$\boxed{L} \rightarrow \boxed{mL} \rightarrow \boxed{cm^3} \rightarrow \boxed{g} \rightarrow \boxed{kg}$$

$$\frac{1\,mL}{0.001\,L} \quad \frac{1\,cm^3}{1\,mL} \quad \frac{19.3\,g}{1\,cm^3} \quad \frac{1\,kg}{1000\,g}$$

Solution:

$$0.82\ L \times \frac{1\,mL}{0.001\,L} \times \frac{1\,cm^3}{1\,mL} \times \frac{19.3\,g}{1\,cm^3} \times \frac{1\,kg}{1000\,g} = 16\ kg$$

SKILLBUILDER 2.17 **Unit Conversion with Equation**

A gold-colored pebble is found in a stream. Its mass is 23.2 mg and its volume is 1.2 mm^3. What is its density in grams per cubic centimeter? Is it gold? (The density of gold is 19.3 g/cm^3.)

Given: m = 23.2 mg

V = 1.2 mm^3

Find: density in g/cm^3

Measurements and Problem Solving

Conversion Factors and Equation:

$$d = \frac{m}{V}$$

1 cm = 10^{-2} m
1 mm = 10^{-3} m
1 mg = 10^{-3} g

Solution Map:

$$\boxed{m, V} \rightarrow \boxed{d}$$

$$d = \frac{m}{V}$$

Solution:

$$d = \frac{m}{V}$$

The equation is already solved for the *find* quantity.
Convert mass from mg to g.

$$m = 23.2 \text{ mg} \times \frac{0.001 \text{ g}}{1 \text{ mg}} = 0.0232 \text{ g}$$

Convert volume from mm^3 to cm^3.

$$V = 1.2 \text{ mm}^3 \times \frac{(0.001 \text{ m})^3}{1 \text{ mm}^3} \times \frac{1 \text{ cm}^3}{(0.01 \text{ m})^3} = 0.0012 \text{ cm}^3$$

Compute density.

$$d = \frac{m}{V} = \frac{0.0232 \text{ g}}{0.0012 \text{ cm}^3} = 19 \frac{\text{g}}{\text{cm}^3}$$

Yes, it is gold.

SELF-TEST QUESTIONS

A. Match the following terms with the phrases below.

conversion factor
decimal part
English system
exponential part

International System (SI)
metric system
prefix multiplier
scientific notation

Chapter 2

        significant figures (digits)   solution map

1. Style of writing numbers that consists of a decimal part and an exponential part
2. Part of scientific notation that is a number usually between 1 and 10
3. Part of scientific notation expressed as 10 raised to an exponent
4. Non-place-holding digits in a reported measurement
5. System of units used in the United States
6. System of units used by most countries in the world
7. Set of standard units agreed upon by scientists throughout the world
8. Multiplier used in the SI system that changes the value of a unit by a power of ten
9. Fraction constructed from two quantities known to be equivalent
10. Visual outline that shows the strategic route required to solve a problem

B. True/False

1. The number 45.6×10^3 is written in scientific notation.
2. The value 3.1060 has 4 significant digits.
3. Exact numbers have an unlimited number of significant figures.
4. To the correct number of significant digits, the sum of $7.2 + 4.8 = 12$.
5. The mass of an object is the measure of the quantity of matter within it.
6. The SI unit of mass is the gram.
7. The SI prefix multiplier milli- means 1000.
8. The SI prefix multiplier for 10^6 is micro.
9. Cubic centimeter (cm^3) is a unit of volume.
10. Density is the ratio of mass to volume.

C. Multiple Choice

1. Express 802,347,894 in scientific notation.
    a) $8.02347894 \times 10^{-9}$
    b) 8.02347894×10^9
    c) 8.02347894×10^8
    d) $8.02,347,894 \times 10^{-8}$

2. Express 3.145×10^{-6} in decimal notation.
    a) 0.0000003145
    b) 0.000003145
    c) 0.003145
    d) 3145000

3. How many significant figures are in the measurement 4.8070 g.
    a) 1
    b) 3
    c) 4
    d) 5

Measurements and Problem Solving

4. Round 273.99876 to four significant figures.
   a) 273.9988
   b) 273.9
   c) 274.0
   d) 274

5. Calculate (238.5 + 456.18) × 0.810 to the correct number of significant figures.
   a) 562.6908
   b) 562.69
   c) 562.7
   d) 563

6. Convert 3.26 g to mg.
   a) 3.26×10^{-6} mg
   b) 3.26×10^{-3} mg
   c) 3.26×10^{3} mg
   d) 3.26×10^{6} mg

7. Convert 8.5×10^{-4} L to mL.
   a) 8.5×10^{-7} mL
   b) 8.5×10^{-3} mL
   c) 8.5×10^{-1} mL
   d) 8.5×10^{2} mL

8. Convert 6.8 mm to in.
   a) 0.97 in
   b) 9.7 in
   c) 2.7 in
   d) 0.27 in

9. Convert 8.44 ft^3 to L.
   a) 1.66 L
   b) 37.0 L
   c) 257 L
   d) 239 L

10. Ethanol has a density of 0.789 g/mL. What is the volume of 80.45 g ethanol?
    a) 63.5 mL
    b) 102 mL
    c) 9.81 mL
    d) 81.24 mL

Chapter 2

D. Crossword Puzzle

ACROSS

1. Ratio of a substance's mass to its volume
6. Standard unit of mass in the SI system
8. Standard unit of length in the SI system
9. Nonnumerical part of a measured value

DOWN

2. Power to which a number is raised
3. Unit of volume
4. Measure of space
5. Quantity of matter
7. Standard unit of time in the SI system

ANSWERS TO SELF-TEST QUESTIONS

A. Matching
1. scientific notation 2. decimal part 3. exponential part 4. significant figures (digits) 5. English system 6. metric system 7. International System (SI) 8. prefix multiplier 9. conversion factor 10. solution map

B. True/False
1. F 2. F 3. T 4. F 5. T 6. F 7. F 8. F 9. T 10. T

C. Multiple Choice
1. c 2. b 3. d 4. c 5. d 6. c 7. c 8. d 9. d 10. b

D. Crossword Puzzle

	1	2	3	4	5	6	7	8	9	10	11
1	¹D	²E	N	S	I	T	Y		³L		⁴V			⁵M	
2		X							⁶K	I	L	O	G	R	A	M
3		P							T		L				S	
4		O							E		U				S	
5		N		⁷S					R		M					
6	⁸M	E	T	E	R						E					
7		N		C												
8		T		O												
9			⁹U	N	I	T										
10				D												

Across:
1. DENSITY
6. KILOGRAM
8. METER
9. UNIT

Down:
1. (none visible)
2. EXPONENT
3. LITER
4. VOLUME
5. MASS
7. SECOND

Chapter 2

Matter and Energy 3

CHAPTER OVERVIEW

Chapter 3 examines the physical and chemical properties of matter. Elements and compounds are introduced. Conservation of mass and energy are discussed. Energy forms and unit conversions are presented. Temperature scales are described. Heat capacity, temperature change, and energy are related.

CHAPTER OBJECTIVES

After reading and studying the text, students should be able to:

1. State the three physical states of matter and their properties.
2. Describe two forms of solid matter.
3. Classify examples of matter as pure substances or mixtures.
4. Classify pure substances as elements or compounds.
5. Classify mixtures as homogeneous or heterogeneous.
6. Classify properties of matter as physical or chemical.
7. Classify changes in matter as physical or chemical.
8. State and apply the law of conservation of mass.
9. State the law of conservation of energy.
10. Differentiate between kinetic energy and potential energy.
11. Convert among energy units of joule, calorie, Calorie and kilowatt-hour.
12. Convert among Fahrenheit, Celsius, and Kelvin temperature scales.
13. Perform calculations relating heat energy to specific heat capacity, mass, and temperature change.

CHAPTER IN REVIEW

- Matter is anything that has mass and occupies space.

- Matter is composed of atoms.

- Atoms bond together to form molecules.

- Matter exists in three physical states: solid, liquid, and gas.

- Solid matter can be amorphous or crystalline.

Chapter 3

- Matter can be classified as a pure substance or a mixture.

- Pure substances can be classified as elements or compounds.

- Elements cannot be broken down into simpler substances.

- Compounds can be broken down into simpler compounds and/or elements.

- Mixtures are composed of two or more different substances whose proportions may vary from one sample to another.

- Mixtures can be classified as homogeneous or heterogeneous.

- Homogeneous mixtures have the same composition throughout.

- Heterogeneous mixtures have two or more regions of different composition.

- Properties of matter can be classified as physical or chemical.

- Physical properties of matter are those that are displayed without a change in composition.

- Chemical properties of matter are those that are displayed by a change in chemical composition.

- Changes in matter can be classified as physical or chemical.

- A physical change is a change in the appearance, but not the composition, of matter.

- A chemical change is a change in the composition of matter.

- Mass is conserved during a physical or chemical change.

- Energy exists in different forms, kinetic energy, electrical energy, and chemical energy.

- Energy can be converted from one form to another; it cannot be created or destroyed.

- The SI unit of energy is the joule. Other units of energy are calorie, Calorie, and kilowatt-hour.

- The temperature of matter is related to the random motion of the atoms and molecules that compose it. As the temperature of matter increases, its atoms and molecules move more rapidly.

Matter and Energy

- Commonly used temperature scales are Fahrenheit, Celsius, and Kelvin.

- The heat capacity of a substance is the quantity of heat energy required to raise the temperature of a given amount of the substance by 1 °C.

- Specific heat capacity is the quantity of heat energy required to raise the temperature of 1 g of a substance by 1 °C. Units of specific heat capacity are J/g °C.

SKILLBUILDER PROBLEMS AND SOLUTIONS

SKILLBUILDER 3.1   **Classifying Matter**

Classify each of the following as a pure substance or mixture. If it is a pure substance, classify it as an element or a compound. If it is a mixture, classify it as homogeneous or heterogeneous.

a) mercury in a thermometer
b) exhaled air
c) minestrone soup
d) sugar

Solution:
a) Mercury is listed in the table of elements. It is a pure substance and an element.
b) Exhaled air consists of several substances, including carbon dioxide and water vapor, and is a mixture. It is uniform throughout, so it is a homogeneous mixture.
c) Minestrone soup contains a number of different substances and is therefore a mixture. There are several different regions within the soup; it is a heterogeneous mixture.
d) Sugar is a pure substance, but it is not listed in the periodic table of elements. It is a compound.

SKILLBUILDER 3.2   **Physical and Chemical Properties**

Determine whether each of the following is a physical or chemical property.

a) the explosiveness of hydrogen gas
b) the bronze color of copper
c) the shiny appearance of silver
d) the ability of dry ice to vaporize without melting

Chapter 3

Solution:
a) Hydrogen gas reacts explosively with oxygen to form water; this is a chemical property.
b) Copper displays its bronze color without changing its composition; this is a physical property.
c) Silver appears shiny without changing its composition; this is a physical property.
d) Dry ice, solid carbon dioxide, is able to convert directly from the solid state to the gas state; this is a physical property.

SKILLBUILDER 3.3   Physical and Chemical Changes

Determine which of the following is a physical or chemical change.

a) copper metal forming a blue solution when it is dropped into colorless nitric acid
b) a passing train flattening a penny placed on a railroad track
c) ice melting into liquid water
d) a match igniting a firework

Solution:
a) Copper metal reacts with nitric acid to form blue copper ions; this is a chemical change.
b) The shape of a penny changes when it is flattened; this is a physical change.
c) When ice melts, water in the solid state converts to water in the liquid state; this is a physical change.
d) When fireworks are ignited a change in chemical composition occurs; this is a chemical change.

SKILLBUILDER 3.4   Conservation of Mass

Suppose 12 g of natural gas combines with 48 g of oxygen in a flame. The chemical change produces 33 g of carbon dioxide and how many grams of water?

Given: 12 g natural gas, 48 g oxygen, 33 g carbon dioxide

Find: g water

Solution:
The sum of the masses of natural gas and oxygen is equal to the sum of the masses of carbon dioxide and water. The mass of water is equal to the sum of the masses of natural gas and oxygen minus the mass of carbon dioxide.

$$12 \text{ g} + 48 \text{ g} = 60 \text{ g}$$

$$60 \text{ g} - 33 \text{ g} = 27 \text{ g}$$

Matter and Energy

SKILLBUILDER 3.5 **Conservation of Energy Units**

The complete combustion of a small wooden match produces approximately 512 cal of heat. How many kilojoules are produced?

Given: 512 cal

Find: kJ

Conversion Factors:

  4.184 J = 1 cal

  1 kJ = 1000 J

Solution Map:

$$\boxed{\text{cal}} \rightarrow \boxed{\text{J}} \rightarrow \boxed{\text{kJ}}$$

$$\frac{4.184 \text{ J}}{1 \text{ cal}} \qquad \frac{1 \text{ kJ}}{1000 \text{ J}}$$

Solution:

Follow the solution map to solve the problem and round to the correct number of significant digits.

$$512 \text{ cal} \times \frac{4.184 \text{ J}}{1 \text{ cal}} \times \frac{1 \text{ kJ}}{1000 \text{ J}} = 2.14 \text{ kJ}$$

SKILLBUILDER PLUS

Convert 2.75×10^4 kJ to calories.

Given: 2.75×10^4 kJ

Find: cal

Conversion Factors:

  1000 J = 1 kJ

  1 cal = 4.184 J

Chapter 3

Solution Map:

$$\boxed{kJ} \rightarrow \boxed{J} \rightarrow \boxed{cal}$$

$$\frac{1000 \text{ J}}{1 \text{ kJ}} \qquad \frac{1 \text{ cal}}{4.184 \text{ J}}$$

Solution:

Follow the solution map to solve the problem and round to the correct number of significant figures.

$$2.75 \times 10^4 \text{ kJ} \times \frac{1000 \text{ J}}{1 \text{ kJ}} \times \frac{1 \text{ cal}}{4.184 \text{ J}} = 6.57 \times 10^6 \text{ cal}$$

SKILLBUILDER 3.6   **Converting between Celsius and Kelvin Temperature Scales**

Convert 358 K to Celsius.

Given: 358 K

Find: °C

Equation:

$$K = °C + 273$$

Solution Map:

$$\boxed{K} \rightarrow \boxed{°C}$$

$$K = °C + 273$$

Solution:
The equation below the arrow shows the relationship between K and °C, but it is not in the correct form. We must solve the equation for °C.

$$K = °C + 273$$

$$°C = K - 273$$

Finally, we substitute the given value for K and compute the answer to the correct number of significant figures.

$$°C = 358 - 273 = 85 \; °C$$

SKILLBUILDER 3.7   **Converting Between Fahrenheit and Celsius Temperature Scales**

Convert 139 °C to Fahrenheit.

Given: 139 °C

Find: °F

Equation:

$$°C = \frac{°F - 32}{1.8}$$

Solution Map:

$$\boxed{°C} \rightarrow \boxed{°F}$$

$$°C = \frac{°F - 32}{1.8}$$

Solution:
The equation beneath the arrow in the solution map shows the relationship between the find quantity (°F) and the given quantity (°C). The numbers 1.8 and 32 are exact in this relationship. The equation must be solved for °F.

$$°C = \frac{°F - 32}{1.8}$$

$$1.8 \; °C = °F - 32$$

$$°F = 1.8 \; (°C) + 32$$

We then substitute into this equation to convert from °C to °F and compute the answer to the correct number of significant figures.

$$°F = 1.8 \; (°C) + 32$$

$$°F = 1.8 \; (139) + 32 = 282 \; °F$$

Chapter 3

SKILLBUILDER 3.8   **Converting Between Fahrenheit and Kelvin Temperature Scales**

Convert −321 °F to kelvin.

Given: −321 °F

Find: K

Equations:
This problem requires two equations: one relating °C and °F and the other relating K and °C.

$$°C = \frac{°F - 32}{1.8}$$

$$K = °C + 273$$

Solution Map:
This conversion requires two steps, one to convert from °F to °C and one to convert from °C to K.

$$\boxed{°F} \rightarrow \boxed{°C} \rightarrow \boxed{K}$$

$$°C = \frac{°F - 32}{1.8} \quad K = °C + 273$$

Solution:
Substitute into the first equation to convert from °F to °C.

$$°C = \frac{°F - 32}{1.8} = \frac{-321 - 32}{1.8} = -196 \ °C$$

We then substitute into the second equation to convert from °C to K.

$$K = °C + 273$$

$$K = -196 + 273 = 77 \ K$$

SKILLBUILDER 3.9   **Relating Heat Energy to Temperature Changes**

You find a copper penny (pre-1982) in the snow and pick it up. How much heat is absorbed by the penny as it warms from the temperature of the snow, −5.0 °C, to the temperature of your body, 37.0 °C? Assume that the penny is pure copper and has a mass of 3.10 g. The specific heat capacity of copper is 0.385 J/g °C.

Given: 3.10 g copper

$$T_i = -5.0 \,°C$$

$$T_f = 37.0 \,°C$$

$$C = 0.385 \text{ J/g °C}$$

Find: q

Equations:

$$q = m \cdot C \cdot \Delta T$$

$$\Delta T = T_f - T_i$$

Solution Map:

$$\boxed{C, m, \Delta T} \;\rightarrow\; \boxed{q}$$

$$q = m \cdot C \cdot \Delta T$$

The necessary quantities for the equation are C, m, and ΔT. The values of these are

$$C = 0.385 \text{ J/g °C}$$

$$m = 3.10 \text{ g}$$

$$\Delta T = T_f - T_i = 37.0 \,°C - (-5.0 \,°C) = 42.0 \,°C$$

Solution:
We then substitute the correct values into the equation, canceling units, and compute the answer to the right number of significant figures.

$$q = m \cdot C \cdot \Delta T$$

$$q = 3.10 \text{ g} \times 0.385 \,\frac{\text{J}}{\text{g °C}} \times 42.0 \,°C = 50.1 \text{ J}$$

Chapter 3

SKILLBUILDER PLUS

The temperature of a lead fishing weight rises from 26 °C to 38 °C as it absorbs 11.3 J of heat. What is the mass of the fishing weight in grams? The specific heat capacity of lead is 0.128 J/g°C.

Given: $q = 11.3$ J
$T_i = 26$ °C
$T_f = 38$ °C
$C = 0.128$ J/g °C

Find: m

Equations: $q = m \cdot C \cdot \Delta T$

$\Delta T = T_f - T_i$

Solution Map:

$$\boxed{q, C, \Delta T} \rightarrow \boxed{m}$$

$$q = m \cdot C \cdot \Delta T$$

The necessary quantities for the equation are q, C, and ΔT. The values of these are

$C = 0.128$ J/g °C

$m = 3.10$ g

$\Delta T = 38$ °C $-$ (26 °C) $= 12$ °C

Solution:

$$q = m \cdot C \cdot \Delta T$$

$$m = \frac{q}{C \cdot \Delta T}$$

Substitute the correct values into the equation, canceling units, and compute the answer to the correct number of significant figures.

$$m = \dfrac{11.3 \text{ J}}{0.128 \dfrac{\text{J}}{\text{g} \, °\text{C}} \times 12 \, °\text{C}} = 7.4 \text{ g}$$

SKILLBUILDER 3.10 — Relating Heat Capacity to Temperature Changes

A 328 g sample of water absorbs 5.78×10^3 J of heat. Find the change in the temperature for the water. If the water is initially at 25.0 °C, what is its final temperature?

Given:

$m = 328$ g water

$q = 5.78 \times 10^3$ J

$T_i = 25.0 \, °\text{C}$

$C = 4.18 \text{ J/g } °\text{C}$

Find: ΔT, T_f

Equations:

$q = m \cdot C \cdot \Delta T$

$\Delta T = T_f - T_i$

Solution Map:

$\boxed{q, m, C} \;\rightarrow\; \boxed{\Delta T} \;\rightarrow\; \boxed{T_f}$

$\quad\quad\quad q = m \cdot C \cdot \Delta T \quad\quad \Delta T = T_f - T_i$

Solution:
Solve the first equation for ΔT. Substitute the correct values into the equation, canceling units, and compute the answer to the correct number of significant figures.

$q = m \cdot C \cdot \Delta T$

$\Delta T = \dfrac{q}{m \cdot C}$

37

Chapter 3

$$\Delta T = \frac{5.78 \times 10^3 \text{ J}}{328 \text{ g} \times 4.18 \frac{\text{J}}{\text{g °C}}} = 4.22 \text{ °C}$$

We solve the second equation for T_f.

$$\Delta T = T_f - T_i$$

$$T_f = \Delta T + T_i$$

Finally, we substitute the correct values into the equation and compute the answer.

$$T_f = \Delta T + T_i = 4.22 \text{ °C} + 25.0 \text{ °C} = 29.2 \text{ °C}$$

SELF-TEST QUESTIONS

A. Match the following terms with the phrases below.

compressible
Calorie (C)
Celsius (°C) scale
chemical change
chemical energy
chemical properties
decanting
distillation
electrical energy
Fahrenheit (°F) scale
filtration
heat capacity
heterogeneous mixture
homogeneous mixture
joule (J)

Kelvin (K) scale
kilowatt-hour (kWh)
kinetic energy
law of conservation of energy
liquid
mixture
perpetual motion machine
physical change
physical properties
potential energy
properties
pure substance
solid
specific heat capacity
volatile

1. State of matter in which atoms or molecules pack close to each other in fixed locations
2. State of matter in which atoms or molecules are packed close together but are free to move around and by each other
3. Substance composed of only one type of atom or molecule
4. Matter composed of two or more different types of atoms or molecules combined in variable proportions
5. Mixture that has two or more regions with different compositions
6. Mixture that has the same composition throughout
7. Properties that a substance displays without changing its composition
8. Properties that a substance displays only through changing its composition
9. Change in matter with no change in composition

10. Change in composition of matter
11. Energy can be neither created nor destroyed
12. Energy associated with motion
13. Energy associated with position
14. Energy associated with the flow of electrical charge
15. Energy associated with potential chemical changes
16. SI unit of energy
17. Unit of energy equal to 1000 calories
18. Unit of energy that appears on electricity bills
19. Temperature scale used commonly in the United States
20. Temperature scale used by scientists that contains negative values
21. Temperature scale used by scientists that has no negative values
22. Quantity of heat energy required to change the temperature of a given amount of substance by 1 °C.
23. Amount of energy required to raise the temperature of 1 g of a substance by 1 °C
24. Able to be forced into less space
25. Pouring off
26. Separation process in which a liquid mixture is heated
27. Process in which a mixture composed of a solid and liquid is poured through filter paper in a funnel
28. Device that supposedly produces energy without the need for energy input
29. Characteristics used to distinguish one substance from another
30. Easily vaporizable

B. True/False

1. Matter is anything that occupies space and has mass.
2. Liquids and solids, but not gases, are matter.
3. Glass and plastic are crystalline solids.
4. Air contains primarily nitrogen and oxygen.
5. Compounds are pure substances.
6. Density is a chemical property.
7. Melting is a physical change.
8. Kinetic energy is associated with motion.
9. Joule is a unit of energy.
10. Absolute zero, 0 K, is the coldest temperature conceivable.

C. Multiple Choice

1. Maple syrup is classified as a(n)
   a) Element
   b) Compound
   c) Homogeneous mixture
   d) Heterogeneous mixture

Chapter 3

2. Which of the following is a chemical property?
   a) The density of lead is 11.4 g/cm³
   b) Butane is a gas at room temperature
   c) Styrene polymerizes to form polystyrene
   d) Acetone evaporates quickly

3. Which of the following is a chemical change?
   a) A newspaper is folded
   b) A newspaper is torn
   c) The ink on a newspaper fades
   d) The ink on a newspaper is smudged

4. Potassium iodide reacts with chlorine to form potassium chloride and iodine. If 7.16 g potassium iodide reacts with 1.53 g chlorine to form 3.22 g potassium chloride, how many grams of iodine form?
   a) 2.41 g
   b) 5.47 g
   c) 8.85 g
   d) 11.91 g

5. Convert 68 °F to K.
   a) 293 K
   b) 338 K
   c) 359 K
   d) 453 K

6. Convert 3.16×10^2 kcal to joules.
   a) 75.5 J
   b) 7.55×10^4 J
   c) 1.32×10^3 J
   d) 1.32×10^6 J

7. Convert 2.34 kWh to kcal.
   a) 2.02×10^3 kcal
   b) 3.52×10^4 kcal
   c) 2.02×10^6 kcal
   d) 3.52×10^7 kcal

8. How many kilojoules of heat are required to raise the temperature of 455 mL of water from 22.4 °C. to 80.0 °C? The specific heat capacity of liquid water is 4.18 J/g °C.
   a) 1.10×10^2 kJ
   b) 2.41×10^2 kJ
   c) 6.27×10^3 kJ
   d) 3.41×10^6 kJ

Matter and Energy

9. When 878 J of heat are absorbed by 125 g of silver, the temperature rises from 25.0 °C to 54.9 °C. What is the specific heat capacity of silver?
   a) 0.128 J/g °C
   b) 0.235 J/g °C
   c) 0.281 J/g °C
   d) 4.26 J/g °C

10. A copper wire weighing 3.16 g absorbs 42.8 J of heat. If its initial temperature is 22.6 °C, what is its final temperature? The specific heat capacity of copper is 0.385 J/g °C.
    a) 27.8 °C
    b) 35.2 °C
    c) 36.1 °C
    d) 57.8 °C

D. Crossword Puzzle

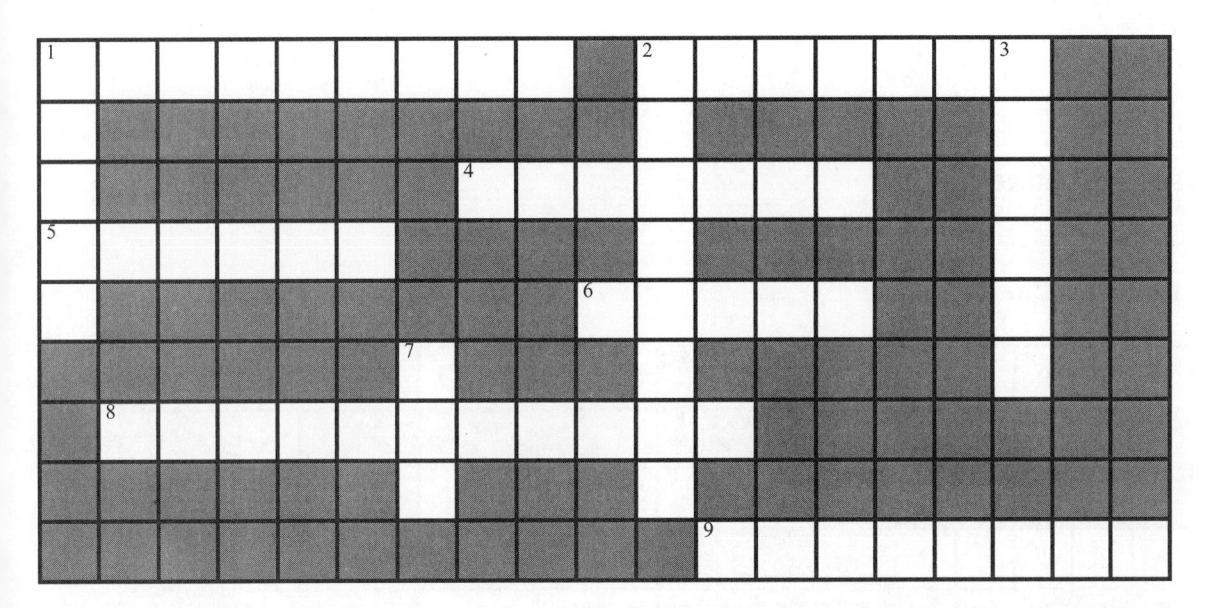

ACROSS

1. Type of solid in which atoms or molecules do not have long-range order
2. Unit of energy equal to 4.184 J
4. Substance that cannot be broken down into simpler substances
5. Anything that has mass and occupies space
6. SI unit of energy
8. Type of solid in which atoms or molecules are arranged in a geometric pattern with long-range, repeating order

DOWN

1. Particles of which all matter is ultimately composed
2. Substance composed of two or more elements in fixed definite proportions
3. Capacity to do work
7. State of matter that is compressible

41

Chapter 3

      9. Cluster of two or more atoms bonded together

ANSWERS TO SELF-TEST QUESTIONS

A. Matching
1. solid 2. liquid 3. pure substance 4. mixture 5. heterogeneous mixture 6. homogeneous mixture 7. physical properties 8. chemical properties 9. physical change 10. chemical change 11. law of conservation of energy 12. kinetic energy 13. potential energy 14. electrical energy 15. chemical energy 16. joule (J) 17. Calorie (C) 18. kilowatt-hour (kWh) 19. Fahrenheit (°F) scale 20. Celsius (°C) scale 21. Kelvin scale 22. heat capacity 23. specific heat capacity 24. compressible 25. decanting 26. distillation 27. filtration 28. perpetual motion machine 29. properties 30. volatile

B. True/False
1. T 2. F 3. F 4. T 5. T 6. F 7. T 8. T 9. T 10. T

C. Multiple Choice
1. c 2. c 3. c 4. b 5. a 6. d 7. a 8. a 9. b 10. d

D. Crossword Puzzle

¹A	M	O	R	P	H	O	U	S		²C	A	L	O	R	I	³E
T										O				N
O					⁴E	L	E	M	E	N	T			E
⁵M	A	T	T	E	R			P					R
S							⁶J	O	U	L	E		G
				⁷G		U					Y
	⁸C	R	Y	S	T	A	L	L	I	N	E
				S				D
							⁹M	O	L	E	C	U	L	E

42

Atoms and Elements

CHAPTER OVERVIEW

Chapter 4 looks at atomic structure from a historical perspective. Properties of protons, neutrons, and electrons are examined. The periodic table is introduced. Isotopes and atomic mass calculations are described.

CHAPTER OBJECTIVES

After reading and studying the text, students should be able to:

1. State the major postulates of Dalton's atomic theory.
2. State the major postulates of Rutherford's nuclear theory of the atom.
3. List the names, symbols, electrical charges, and relative masses of the three major subatomic particles.
4. Summarize the nature of electrical charge in terms of attraction and repulsion.
5. Use the periodic table to determine the atomic number and the atomic mass of an element.
6. Use the periodic table to classify elements as metals, nonmetals, or metalloids.
7. Use the periodic table to classify an element as a main-group element or a transition metal.
8. Use the periodic table to identify alkali metals, alkaline earth metals, halogens, and noble gases.
9. Determine the charge of a main group ion from its position in the periodic table.
10. Determine the number of protons and electrons in an ion, given its formula.
11. Determine the atomic number, mass number, and symbol of an isotope, given the identity of the element and the number of neutrons. Write the symbol of an isotope, given its number of protons and neutrons.
12. Calculate the atomic mass of an element, given the mass and percent natural abundance of each isotope.

CHAPTER IN REVIEW

- All matter is composed ultimately of atoms. Properties of atoms determine the properties of matter.

- An atom is the smallest identifiable unit of an element.

Chapter 4

- Dalton's atomic theory states that matter is composed of atoms, that atoms of a given element have unique properties that distinguish them from atoms of other elements, and that atoms combine in simple whole-number ratios to form compounds.

- J. J. Thomson, an English physicist, discovered the electron. He discovered that electrons were negatively charged and that they were much smaller and lighter than atoms.

- Electrons are negatively charged particles that are a fundamental part of the atom.

- Ernest Rutherford proposed the nuclear theory of the atom based on experiments in which he bombarded ultrathin sheets of gold foil with alpha particles.

- The nuclear theory of the atom states that most of an atom's mass and all of its positive charge are contained in a small core called the nucleus. Most of the volume of the atom is empty space occupied by tiny, negatively charged electrons. There are as many negatively charged electrons outside the nucleus as there are positively charged particles (protons) inside the nucleus. The atom is electrically neutral.

- The nucleus of the atom contains positively charged protons and neutral particles called neutrons. Protons and neutrons have similar masses, approximately 1 amu.

- Electrical charge is a fundamental property of protons and electrons. Protons are positively charged and electrons are negatively charged.

- Like electrical charges repel; opposite electrical charges attract.

- The number of protons in an atom's nucleus defines the element. This number is called the atomic number and is symbolized by the letter Z.

- The periodic table lists all known elements according to their atomic number. Each element is represented by a chemical symbol, a one- or two-letter abbreviation.

- Dmitri Mendeleev proposed the periodic law and arranged early versions of the periodic table. The periodic law states that "When the elements are arranged in order of increasing relative mass, certain sets of properties recur periodically."

- Metals tend toward the left of the periodic table. Metals are good conductors of heat and electricity, malleable, ductile, often shiny, and tend to lose electrons when they undergo chemical change.

- Nonmetals tend toward the right of the periodic table. They tend to be poor conductors of heat and electricity and tend to gain electrons when they undergo chemical change.

- Metalloids fall along the zigzag diagonal line of the periodic table that divides metals and nonmetals. Metalloids are semiconductors of electricity.

Atoms and Elements

- The periodic table can be broadly divided into main group elements, whose properties tend to be more predictable based on their position in the periodic table, and transition elements, whose properties tend to be less predictable based on their position in the periodic table.

- Columns in the periodic table are called groups or families. The Group 1A elements are called alkali metals. The Group 2A elements are called alkaline earth metals. The Group 7A elements are called halogens. The Group 8A elements are called noble gases.

- Atoms often gain or lose electrons to form ions. Cations are positively charged ions; anions are negatively charged ions.

- The Group 1A metals tend to lose 1 electron to form a +1 ion. The Group 2A elements tend to lose 2 electrons to form a +2 ion. The Group 7A elements tend to gain 1 electron to form –1 ions.

- Isotopes are atoms of the same element with different numbers of neutrons. The percent natural abundance of an isotope is the relative amount of the isotope in a naturally occurring sample of the element.

- The mass number of an isotope is the sum of the number of protons and number of neutrons and is represented by the symbol A.

- The atomic mass of each element is listed beneath the symbol of the element on the periodic table. Atomic mass is an average calculated by multiplying the fraction of each isotope by its isotopic mass, then summing the results.

SKILLBUILDER PROBLEMS AND SOLUTIONS

SKILLBUILDER 4.1 Atomic Number, Atomic Symbol, and Element Name

Find the name and atomic number for each of the following elements.

a) Na
b) Ni
c) P
d) Ta

Chapter 4

Solution:

	Element	Atomic Number
a)	sodium	11
b)	nickel	28
c)	phosphorus	15
d)	tantalum	73

SKILLBUILDER 4.2   Classifying Elements as Metals, Nonmetals, or Metalloids

Classify each of the following elements as a metal, nonmetal, or metalloid.

a) S
b) Cl
c) Ti
d) Sb

Solution:
a) Sulfur is on the right-hand side of the periodic table; it is a nonmetal.
b) Chlorine is on the right-hand side of the periodic table; it is a nonmetal.
c) Titanium is on the left-hand side of the periodic table; it is a metal.
d) Antimony is in the middle-right section of the periodic table in the area of the metalloids; it is a metalloid.

SKILLBUILDER 4.3   Groups and Families of Elements

To which group or family of elements does each of the following elements belong?

a) Li
b) B
c) I
d) Ar

Solution:
a) Lithium is in Group 1A; it is an alkali metal.
b) Boron is in Group 3A.
c) Iodine is in Group 7A; it is a halogen.
d) Argon is in Group 8A; it is a noble gas.

Atoms and Elements

| SKILLBUILDER 4.4 | **Determining Ion Charge from Numbers of Protons and Electrons** |

Determine the charge of each of the following ions.

a) a nickel ion with 26 electrons
b) a bromine ion with 36 electrons
c) a phosphorus ion with 18 electrons

Solution:
To determine the charge of each ion, we use the ion charge equation.

$$\text{Ion charge} = \#\,p - \#\,e^-$$

The number of electrons is given in the problem. The number of protons is obtained from the element's atomic number in the periodic table.

a) Nickel with atomic number 28 has 28 protons in its nucleus.

$$\text{Ion charge} = 28 - 26 = +2 \quad (\text{Ni}^{2+})$$

b) Bromine with atomic number 35 has 35 protons in its nucleus.

$$\text{Ion charge} = 35 - 36 = -1 \quad (\text{Br}^-)$$

c) Phosphorus with atomic number 15 has 15 protons in its nucleus.

$$\text{Ion charge} = 15 - 18 = -3 \quad (\text{P}^{3-})$$

| SKILLBUILDER 4.5 | **Determining the Number of Protons and Electrons in an Ion** |

Find the number of protons and electrons in the S^{2-} ion.

Solution:
From the periodic table, we find that the atomic number for sulfur is 16, so sulfur has 16 protons. The number of electrons can be found using the ion charge equation.

$$\text{Ion charge} = \#\,p - \#\,e^-$$
$$-2 = 16 - \#\,e^-$$
$$\#\,e^- = 16 - (-2) = 18$$

Chapter 4

SKILLBUILDER 4.6 — Charge of Ions from Position in Periodic Table

Based on their position in the periodic table, what ions will potassium and sulfur tend to form?

Solution:
Since potassium is in Group 1A, it will tend to form an ion with a +1 charge (K^+). Since sulfur is in Group 6A, it will tend to form an ion with a −2 charge (S^{2-}).

SKILLBUILDER 4.7 — Atomic Numbers, Mass Numbers, and Isotope Symbols

What are the atomic number, mass number, and symbol for the chlorine isotope with 18 neutrons?

Solution:
From the periodic table, we find that the atomic number (Z) of chlorine is 17, so chlorine atoms have 17 protons. The mass number (A) for the isotope with 18 neutrons is the sum of the number of protons and the number of neutrons.

$$A = 17 + 18 = 35$$

So, $Z = 17$, $A = 35$, and the symbol for the isotope chlorine-35 is $^{35}_{17}Cl$.

SKILLBUILDER 4.8 — Numbers of Protons and Neutrons from Isotope Symbols

How many protons and neutrons are in the following potassium isotope?

$$^{39}_{19}K$$

Solution:
The number of protons is equal to Z (lower left number).

$$\# p = Z = 19$$

The number of neutrons is equal to A (upper left number) − Z (lower left number).

$$\# n = 39 - 19 = 20$$

SKILLBUILDER 4.9 — Calculating Atomic Mass

Magnesium has three naturally occurring isotopes with masses of 23.99 amu, 24.99 amu, and 25.98 amu and natural abundances of 78.99%, 10.00%, and 11.01%, respectively. Calculate the atomic mass of magnesium.

Atoms and Elements

Solution:
Convert the percent natural abundances into decimal form by dividing by 100.

$$\frac{78.99}{100} = 0.7899$$

$$\frac{10.00}{100} = 0.1000$$

$$\frac{11.01}{100} = 0.1101$$

Use the fractional abundances and the atomic masses of the isotopes to compute the atomic mass.

Atomic mass = (0.7899 × 23.99 amu) + (0.1000 × 24.99 amu) + (0.1101 × 25.98 amu)
= 18.9497 amu + 2.499 amu + 2.8603 amu
= 24.31 amu

SELF-TEST QUESTIONS

A. Match the following terms with the phrases below.

- alkali metals
- alkaline earth metals
- atomic mass
- atomic mass unit (amu)
- atomic number (Z)
- chemical symbol
- family or group (of elements)
- main-group elements
- mass number (A)
- noble gases
- nuclear radiation
- nuclear theory of the atom
- percent natural abundance
- periodic law
- periodic table
- radioactive
- transition metals

1. Organized chart of the elements
2. Theory that states in part that most of the volume of the atom is empty space
3. One-twelfth of the mass of a carbon atom containing six protons and six neutrons
4. Number of protons in the nucleus of an atom
5. One- or two-letter abbreviation for an element
6. When the elements are arranged in order of increasing relative mass, certain sets of properties recur periodically
7. Elements whose properties tend to be more predictable based on their position in the periodic table
8. Elements whose properties tend to be less predictable based simply on their position in the periodic table
9. Column of elements in the periodic table

10. Group 8A elements
11. Group 1A elements
12. Group 2A elements
13. Relative amount of an isotope in a naturally occurring sample of a given element
14. Sum of the number of protons and neutrons in an atom
15. Subatomic particles emitted from unstable nuclei
16. Average mass of the atoms that compose an element
17. Characteristic of unstable nuclei that emit radiation

B. True/False

1. The nucleus of an atom is neutral.
2. Protons and electrons are located outside the nucleus.
3. Most of the atom's mass is concentrated in the nucleus.
4. Carbon atoms have 12 protons.
5. Group 2A elements are called alkali metals.
6. Cations are positively charged ions.
7. Halogens tend to form anions with a –1 charge.
8. Isotopes are atoms of the same element with different numbers of electrons.
9. Phosphorus-31 has 16 protons.
10. All chlorine atoms have the same mass.

C. Multiple Choice

1. Which subatomic particle has a +1 charge?
   a) Electron
   b) Neutron
   c) Proton
   d) Nucleus

2. The atomic number of sulfur is
   a) 14
   b) 16
   c) 34
   d) 50

3. The atomic symbol for copper is
   a) Ca
   b) Co
   c) Cr
   d) Cu

Atoms and Elements

4. What is the name of the element symbolized as Sn?
   a) Silicon
   b) Tin
   c) Strontium
   d) Sodium

5. Which of the following elements is a nonmetal?
   a) Lithium
   b) Nickel
   c) Boron
   d) Neon

6. Which of the following is a halogen?
   a) Oxygen
   b) Nitrogen
   c) Chlorine
   d) Argon

7. Give the number of protons and electrons in a Ag^+ ion.
   a) 46 protons, 47 electrons
   b) 47 protons, 47 electrons
   c) 47 protons, 48 electrons
   d) 47 protons, 46 electrons

8. Give the number of protons and electrons in an O^{2-} ion.
   a) 8 protons, 6 electrons
   b) 8 protons, 10 electrons
   c) 6 protons, 4 electrons
   d) 10 protons, 8 electrons

9. What is the isotopic symbol for iodine-131?
   a) $^{53}_{131}I$
   b) $^{131}_{53}I$
   c) $^{184}_{53}I$
   d) $^{78}_{53}I$

10. Copper has two naturally occurring isotopes: copper-63 with mass 62.9298 amu and a natural abundance of 69.17% and copper-65 with mass 64.9278 amu and a natural abundance of 30.83%. Calculate the atomic mass of copper.
    a) 63.93 amu
    b) 63.55 amu
    c) 64.00 amu
    d) 64.31 amu

Chapter 4

D. Crossword Puzzle

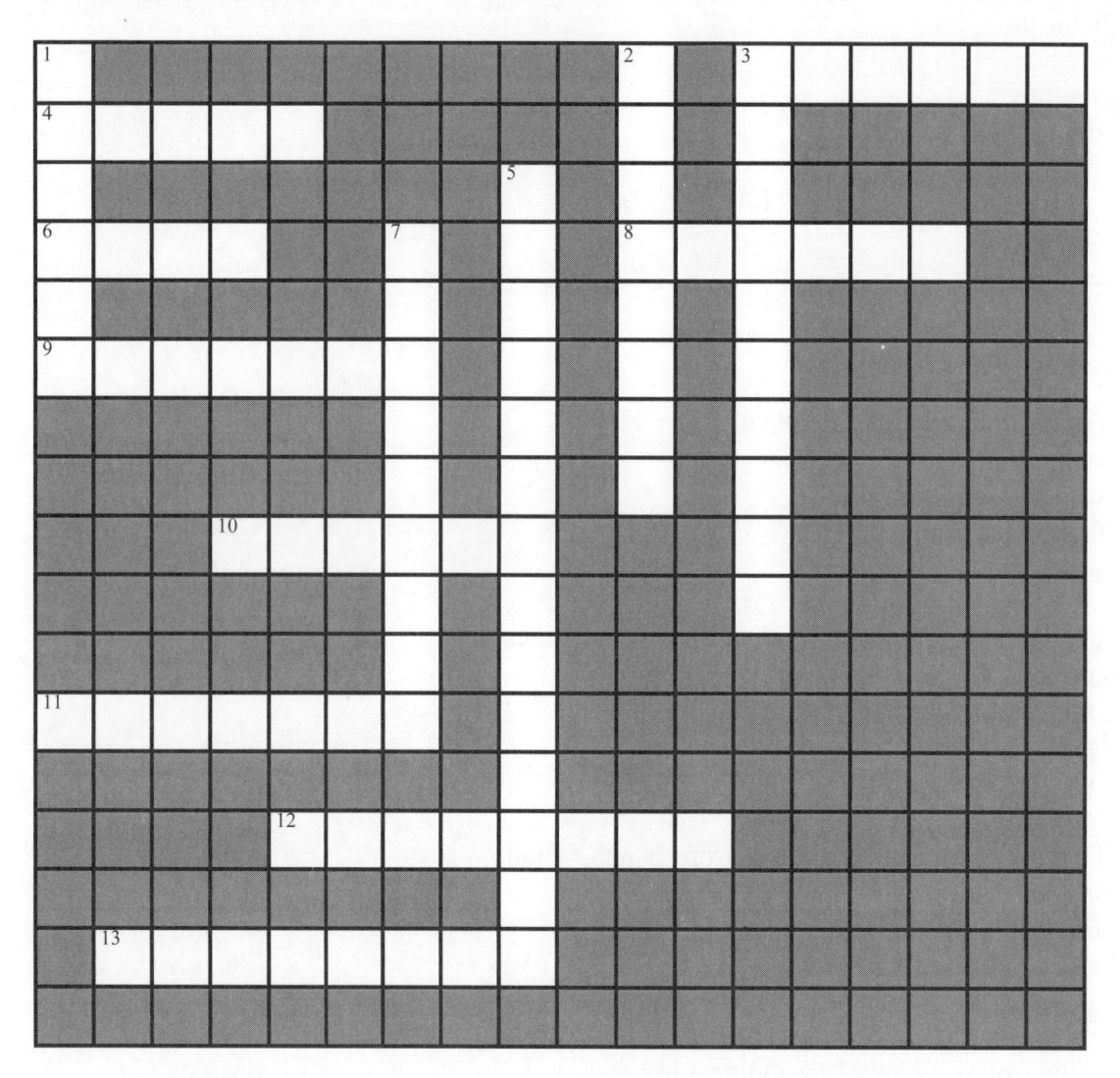

ACROSS

3. Elements that are often shiny, good conductors of heat and electricity
4. Negatively charged ion
6. Charged particles that form when atoms gain or lose electrons
8. Fundamental property of protons that is equal in magnitude but opposite in sign to that of electrons
9. Neutral particle within the nucleus of an atom
10. Positively charged fundamental part of the atom
11. Small core within the atom

DOWN

1. Positively charged ion
2. Negatively charged fundamental part of the atom
3. Elements that fall in the middle-right of the periodic table on the zigzag line that divides metals and nonmetals
5. Elements with intermediate electrical conductivity, which can be changed and controlled
7. Elements that tend to be poor conductors of heat and electricity and tend to gain electrons when they undergo chemical changes

containing most of the its mass and all of its positive charge
12. Atoms with the same number of protons but different numbers of neutrons
13. Group 7A elements

ANSWERS TO SELF-TEST QUESTIONS

A. Matching
1. periodic table 2. nuclear theory of the atom 3. atomic mass unit (amu) 4. atomic number (Z) 5. chemical symbol 6. periodic law 7. main-group elements 8. transition metals 9. family or group (of elements) 10. noble gases 11. alkali metals 12. alkaline earth metals 13. percent natural abundance 14. mass number (A) 15. nuclear radiation 16. atomic mass 17. radioactive

B. True/False
1. F 2. F 3. T 4. F 5. F 6. T 7. T 8. F 9. F 10. F

C. Multiple Choice
1. c 2. b 3. d 4. b 5. d 6. c 7. d 8. b 9. b 10. b

Chapter 4

D. Crossword Puzzle

	1	2	3	4	5	6	7	8	9	10	11	12	13	14	15
1	¹C								²E		³M	E	T	A	L	S
2	⁴A	N	I	O	N				L		E					
3	T							⁵S	E		T					
4	⁶I	O	N	S		⁷N	E	⁸C	H	A	R	G	E			
5	O					O		M	T		L					
6	⁹N	E	U	T	R	O	N	I		R	L					
7						M		C		O	O					
8						E		O		N	I					
9				¹⁰P	R	O	T	O	N		D					
10|					A		D			S					
11|					L		U								
12| ¹¹N	U	C	L	E	U	S		C							
13|								T							
14|			¹²I	S	O	T	O	P	E	S					
15|								R							
16|	¹³H	A	L	O	G	E	N	S							

54

Molecules and Compounds   5

CHAPTER OVERVIEW

Chapter 5 focuses on formula writing and nomenclature. Names and formulas of Type I and Type II ionic compounds, molecular compounds, and acids are discussed.

CHAPTER OBJECTIVES

After reading and studying the text, students should be able to:

1. State and apply the law of constant composition.
2. Write a chemical formula given the number of atoms of each element in the compound.
3. Determine the number of each type of atom in a chemical formula.
4. Classify substances as atomic elements, molecular elements, molecular compounds, or ionic compounds.
5. Write formulas for ionic compounds.
6. Name Type I and Type II binary ionic compounds.
7. Name ionic compounds containing a polyatomic ion.
8. Name molecular compounds.
9. Name binary acids and oxyacids.
10. Calculate the formula mass of a molecule or formula unit.

CHAPTER IN REVIEW

- Joseph Proust was the first chemist to formally state the idea that elements combine in fixed proportions to form compounds.

- The law of constant composition states that all samples of a given compound have the same proportions of their constituent elements.

- Chemical formulas indicate the elements present in the compound and the relative number of atoms of each.

- Subscripts in a chemical formula represent the relative number of each type of atom in a chemical compound; they do not change for a given compound.

- Chemical formulas normally list the most metallic element first.

Chapter 5

- Polyatomic ions are groups of atoms with an overall charge.

- When there is more than one polyatomic ion in a chemical formula, the polyatomic ion is set off in parentheses. The subscript outside the parentheses indicates the number of polyatomic ions in the formula.

- Elements may be atomic or molecular. Atomic elements exist with single atoms as the basic unit. Molecular elements exist as diatomic molecules in which two atoms are bonded together.

- Compounds may be molecular or ionic. Molecular compounds exist with molecules as the basic unit. Molecules consist of two or more different nonmetal atoms bonded together. Ionic compounds are formed between a metal and a nonmetal. The basic unit of an ionic compound is the formula unit.

- Ionic compounds contain positive cations and negative anions. The sum of the positive charges of the cations is equal to the sum of the negative charges of the anions.

- Ionic compounds can be categorized as Type I or Type II. Type I compounds contain a metal that always forms a cation of the same charge. Type II compounds contain a metal that forms cations of different charges. Most transition metals form Type II compounds.

- Binary compounds contain only two different elements.

- To name Type I binary compounds, name the metal, then give the base name of the nonmetal with an *–ide* suffix.

- To name Type II binary compounds, name the metal, state the charge of the metal cation (written in Roman numerals in parentheses), then give the base name of the nonmetal with an *–ide* suffix.

- Ionic compounds containing a polyatomic ion are named the same way as other ionic compounds except that the name of the polyatomic ion is used where it occurs.

- Most polyatomic ions are oxyanions, anions that contain oxygen. Series of oxyanions are named systematically according to the number of oxygen atoms in the ion.

- Molecular compounds are named using prefixes that indicate the number of atoms of each element present in the molecule. State the prefix and the name of the first element followed by the prefix and base name of the second element with an *-ide* suffix. If there is only one atom of the first element in the formula, the prefix *mono-* is normally omitted.

- Acids are molecular compounds that dissolve in water to form H^+ ions. Binary acids contain only hydrogen and a nonmetal; oxyacids contain hydrogen, a nonmetal, and oxygen.

Molecules and Compounds

- Binary acid names begin with *hydro-*, followed by the base name of the nonmetal with an *–ic* suffix, and then the word *acid*.

- Oxyacids are categorized according to the name of the corresponding oxyanion. If the oxyanion name ends in *–ate*, state the nonmetal stem of the oxyanion with an *–ic* suffix, followed by the word *acid*. If the oxyanion name ends in *–ite*, state the nonmetal stem of the oxyanion with an *–ous* suffix, followed by the word *acid*.

- The formula mass of a compound is the average mass of the molecules or formula units that compose the compound.

- The formula mass of a compound is calculated by adding the atomic masses of all the atoms in the chemical formula.

SKILLBUILDER PROBLEMS AND SOLUTIONS

SKILLBUILDER 5.1   Constant Composition of Compounds

Two samples of carbon monoxide, obtained from different sources, were decomposed into their constituent elements. One sample produced 4.3 g of oxygen and 3.2 g of carbon, and the other sample produced 7.5 g of oxygen and 5.6 g of carbon. Are these results consistent with the law of constant composition?

Solution:
To show this, compute the mass ratio of one element to the other for both samples by dividing the larger mass by the smaller one.
For the first sample:

$$\frac{\text{Mass oxygen}}{\text{Mass carbon}} = \frac{4.3 \text{ g}}{3.2 \text{ g}} = 1.3$$

For the second sample:

$$\frac{\text{Mass oxygen}}{\text{Mass carbon}} = \frac{7.5 \text{ g}}{5.6 \text{ g}} = 1.3$$

Since the ratios are the same for the two samples, these results are consistent with the law of constant composition.

Chapter 5

| SKILLBUILDER 5.2 | Writing Chemical Formulas |

Write a chemical formula for each of the following.

a) the compound containing two silver atoms to every sulfur atom.
b) the compound containing two nitrogen atoms to every oxygen atom.
c) the compound containing two oxygen atoms to every titanium atom.

Solution:
a) Since silver is a metal, it is listed first. The formula is Ag_2S.
b) Since nitrogen is to the left of oxygen on the periodic table and since it occurs before oxygen in Table 5.1 in the text, it is listed first. The formula is N_2O.
c) Since titanium is a metal, it is listed first. The formula is TiO_2.

| SKILLBUILDER 5.3 | Total Number of Each Type of Atom in a Chemical Formula |

Determine the number of each type of atom in K_2SO_4.

Solution:
K: There are 2 K atoms as indicated by the subscript 2.
S: There is 1 S atom as indicated by the implied subscript 1.
O: There are 4 O atoms as indicated by the subscript 4.

| SKILLBUILDER PLUS |

Determine the number of each type of atom in $Al_2(SO_4)_3$.

Solution:
K: There are 2 Al atoms as indicated by the subscript 2.
S: There are 3 S atoms as indicated by multiplying the subscript outside of the parentheses (3) by the subscript for S inside the parentheses, which is 1 (implied).
O: There are 12 O atoms as indicated by multiplying the subscript outside of the parentheses (3) by the subscript for O inside the parentheses, which is 4.

| SKILLBUILDER 5.4 | Classifying Substances as Atomic Elements, Molecular Elements, Molecular Compounds, or Ionic Compounds |

Classify each of the following substances as an atomic element, molecular element, molecular compound, or ionic compound.

a) chlorine
b) NO

c) Au
d) Na$_2$O
e) CrCl$_3$

Solution:
a) Chlorine is an element that occurs as diatomic molecules; therefore, it is a molecular element.
b) NO is a compound composed of two nonmetals; therefore, it is a molecular compound.
c) Gold is an element that does not occur as diatomic molecules; therefore, it is an atomic element.
d) Na$_2$O is a compound composed of a metal (left side of periodic table) and nonmetal (right side of periodic table); therefore, it is an ionic compound.
e) CrCl$_3$ is a compound composed of a metal (left side of periodic table) and nonmetal (right side of periodic table); therefore, it is an ionic compound.

SKILLBUILDER 5.5   Writing Formulas for Ionic Compounds

Write a formula for the compound formed from cesium and oxygen.

Solution:
We first write the symbol for each ion along with its appropriate charge from its group number in the periodic table.

$$Cs^+ \quad O^{2-}$$

We then make the magnitude of each ion's charge become the subscript for the other ion.
$$Cs^+ \quad O^{2-} \quad \text{becomes } Cs_2O$$

There is no reducing of subscripts necessary in this case. Finally, we check to see that the sum of the charges of the cations ($+1 + 1 = +2$) exactly cancels the sum of the charges of the anion (-2). The correct formula is Cs$_2$O.

SKILLBUILDER 5.6   Writing Formulas for Ionic Compounds

Write a formula for the compound formed from aluminum and nitrogen.

Solution:
We first write the symbol for each ion along with its appropriate charge from its group number in the periodic table.

$$Al^{3+} \quad N^{3-}$$

We then make the magnitude of each ion's charge become the subscript for the other ion.

$$Al^{3+} \quad N^{3-} \quad \text{becomes } Al_3N_3$$

Chapter 5

To reduce the subscripts, we divide both subscripts by 3.

$$Al_3N_3 \div 3 = AlN$$

Finally, we check to see that the sum of the charges of the cations (+3) exactly cancels the sum of the charges of the anions (−3). The correct formula is AlN.

SKILLBUILDER 5.7   Writing Formulas for Ionic Compounds

Write a formula for the compound that forms from calcium and bromine.

Solution:
We first write the symbol for each ion along with its appropriate charge from its group number in the periodic table.

$$Ca^{2+} \quad Br^-$$

We then make the magnitude of each ion's charge become the subscript for the other ion.

$$Ca^{2+} \quad Br^- \quad \text{becomes } CaBr_2$$

There is no reducing of subscripts necessary in this case. Finally, we check to see that the sum of the charges of the cations (+2) exactly cancels the sum of the charges of the anion (−1 + (−1) = −2). The correct formula is $CaBr_2$.

SKILLBUILDER 5.8   Naming Type I Ionic Compounds

Give the name for the compound KBr.

Solution:
The cation is potassium. The anion is bromine, which becomes *bromide*. The correct name is *potassium bromide*.

SKILLBUILDER PLUS

Give the name for the compound Zn_3N_2.

Solution:
The cation is zinc. The anion is nitrogen, which becomes *nitride*. The correct name is *zinc nitride*.

Molecules and Compounds

SKILLBUILDER 5.9   **Naming Type II Ionic Compounds**

Give the name for the compound PbO.

Solution:
The name for PbO consists of the name of the cation, *lead*, followed by the charge of the cation in parentheses *(II)*, followed by the base name of the anion, *ox-*, with the ending *–ide*. The full name is *lead(II) oxide*. We know the charge on lead is +2 because the charge on O is –2 and there is one O^{2-} anion.

    PbO   lead(II) oxide

SKILLBUILDER 5.10   **Naming Ionic Compounds That Contain a Polyatomic Ion**

Give the name for the compound $Mn(NO_3)_2$.

Solution:
The name for $Mn(NO_3)_2$ consists of the name of the cation, *manganese*, followed by the charge of the cation in parentheses *(II)*, followed by the name of the polyatomic ion, *nitrate*. The full name is *manganese(II) nitrate*. We know the charge on manganese is +2 because the charge on NO_3^- is –1 and there are two NO_3^- anions.

    $Mn(NO_3)_2$ manganese(II) nitrate

SKILLBUILDER 5.11   **Naming Molecular Compounds**

Name the compound N_2O_4.

Solution:
The name of the compound is the name of the first element, *nitrogen*, prefixed by *di-* to indicate two, followed by the base name of the second element, *-ox-*, prefixed by *tetra-* to indicate four, and the suffix *-ide*. The entire name is *dinitrogen tetroxide*.

SKILLBUILDER 5.12   **Naming Binary Acids**

Give the name of HI.

Solution:
The base name of I is *-iod-* so the name is *hydroiodic acid*.

Chapter 5

| SKILLBUILDER 5.13 | Naming Oxyacids |

Give the name of HNO$_2$.

Solution:
The oxyanion is nitrite, which ends in *-ite*; therefore, the name of the acid is *nitrous acid*.

| SKILLBUILDER 5.15 | Calculating Formula Masses |

Calculate the formula mass of dinitrogen monoxide, N$_2$O, also called laughing gas.

Solution:
To find the formula mass, we sum the atomic masses of each atom in the chemical formula

$$\begin{aligned}\text{Formula mass} &= 2 \times (\text{atomic mass N}) + 1 \times (\text{atomic mass O}) \\ &= 2(14.01 \text{ amu}) + 16.00 \text{ amu} \\ &= 44.02 \text{ amu}\end{aligned}$$

SELF-TEST QUESTIONS

A. Match the following terms with the phrases below.

    atomic element   law of constant composition
    chemical formula   molecular element
    formula mass   Type I compound
    formula unit   Type II compound

1. All samples of a given compound have the same proportions of their constituent elements
2. Representation of a compound that indicates the elements present in the compound and the relative number of atoms of each
3. Element that exists in nature with single atoms as the basic unit
4. Element that does not normally exist in nature with single atoms as the basic unit
5. Basic unit of an ionic compound
6. Ionic compound containing a metal that always forms a cation with the same charge
7. Ionic compound containing a metal that forms cations of different charges
8. Average mass of the molecules or formula units that compose a compound

B. True/False

1. The formula Ca$_3$(PO$_4$)$_2$ shows a total of 6 oxygen atoms.
2. The element chlorine exists in nature as a diatomic molecule.
3. The basic unit of an ionic compound is the formula unit.
4. The formula for potassium oxide is KO.

Molecules and Compounds

5. The formula for iron(III) bromide is Fe₃Br.
6. The name for the compound NiCl₂ is nickel(II) chloride.
7. The carbonate ion has a charge of –1.
8. The formula for carbon tetrachloride is CCl₄.
9. Hydrochloric acid is a binary acid.
10. The formula for sulfuric acid is H₂SO₃.

C. Multiple Choice

1. How many oxygen atoms are in the formula Mg(HCO₃)₂?
   a) 2
   b) 3
   c) 5
   d) 6

2. Which of the following is a molecular element?
   a) aluminum
   b) helium
   c) hydrogen
   d) calcium

3. Which of the following compounds is ionic?
   a) NH₃
   b) SO₂
   c) K₂O
   d) CF₄

4. Write the formula for an ionic compound composed of calcium and phosphorus.
   a) CaP
   b) Ca₂P
   c) Ca₃P
   d) Ca₃P₂

5. Give the name of Na₂SO₄.
   a) Sodium sulfuroxygen
   b) Sodium sulfite
   c) Sodium sulfate
   d) Sodium sulfide

6. Write the formula for copper(II) bromide.
   a) Cu₂Br
   b) Cu₂Br₂
   c) CuBr
   d) CuBr₂

Chapter 5

7. Write the formula for diiodine pentaoxide.
    a) I₂O₅
    b) I₂O₄
    c) I₅O₂
    d) (IO₅)₂

8. Give the name of HClO₃.
    a) Hydrochloric acid
    b) Hypochlorous acid
    c) Chlorous acid
    d) Chloric acid

9. Give the name of CuO.
    a) copper oxide
    b) copper monoxide
    c) copper(I) oxide
    d) copper(II) oxide

10. Calculate the formula mass of sodium carbonate, Na₂CO₃.
    a) 51.00 amu
    b) 83.00 amu
    c) 105.99 amu
    d) 130.01 amu

D. Crossword Puzzle

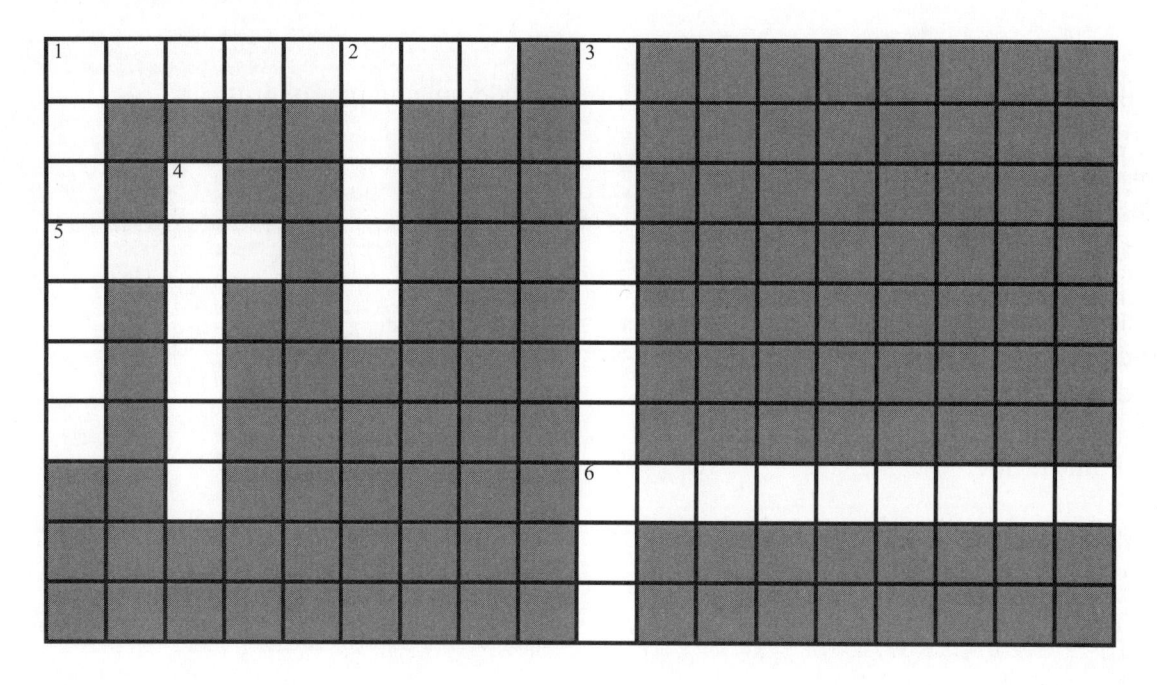

Molecules and Compounds

ACROSS

1. Anion containing oxygen
5. Molecular compound that dissolves in water to form H+ ions
6. Type of compound formed between two or more nonmetals

DOWN

1. Acid that contains hydrogen, a nonmetal, and oxygen
2. Compound formed between a metal and one or more nonmetals
3. Type of ion composed of a group of atoms with an overall charge
4. Type of acid that contains only hydrogen and a nonmetal

ANSWERS TO SELF-TEST QUESTIONS

A. Matching
1. law of constant composition 2. chemical formula 3. atomic element 4. molecular element 5. formula unit 6. Type I compound 7. Type II compound 8. formula mass

B. True/False
1. F 2. T 3. T 4. F 5. F 6. T 7. F 8. T 9. T 10. F

C. Multiple Choice
1. d 2. c 3. c 4. d 5. c 6. d 7. a 8. d 9. d 10. c

D. Crossword Puzzle

¹O	X	Y	A	N	²I	O	N		³P						
X					O				O						
Y		⁴B			N				L						
⁵A	C	I	D		I				Y						
C		N			C				A						
I		A							T						
D		R							O						
		Y							⁶M	O	L	E	C	U	L	A	R
									I						
									C						

65

Chapter 5

Chemical Composition

6

CHAPTER OVERVIEW

Chapter 6 introduces the mole. Mole-to-mole and mole-to-mass conversions are explained. Calculations of empirical and molecular formulas from reaction data and percent composition are described.

CHAPTER OBJECTIVES

After reading and studying the text, students should be able to:

1. Convert between moles and number of atoms.
2. Convert between grams and moles.
3. Convert between grams and number of atoms.
4. Calculate the molar mass of a compound.
5. Convert between mass and number of molecules or formula units.
6. Convert between moles of a compound and moles of a constituent element.
7. Convert between grams of a compound and grams of a constituent element.
8. Calculate mass percent composition of a compound from experimental data.
9. Use mass percent composition to convert between the mass of a compound and the mass of a constituent element.
10. Calculate the mass percent composition of a compound from its chemical formula.
11. Calculate the empirical formula of a compound from experimental data.
12. Calculate the molecular formula of a compound from its empirical formula and molar mass.

CHAPTER IN REVIEW

- The mole is a convenient number to use when dealing with atoms or molecules. The value of 1 mole is 6.022×10^{23}. This number is also called Avogadro's number.

- Avogadro's number is used to convert between number of atoms, molecules, or formula units and number of moles of atoms, molecules, or formula units.

- The molar mass of an element is the mass of one mole of atoms of that element. The value of the molar mass in grams per mole is numerically equivalent to the element's atomic mass in atomic mass units.

- The molar mass of an element is used to convert between the mass of the element and the number of moles of the element.

- To convert from the mass of an element to the number of atoms, use the molar mass of the element to convert from grams to moles, then use Avogadro's number to convert from moles to number of atoms.

- The molar mass of a compound is the mass of one mole of molecules or formula units of that compound. The value of the molar mass in grams per mole is numerically equivalent to the formula mass of the compound in atomic mass units.

- The molar mass of a compound is used to convert between the mass of the compound and the number of moles of the compound.

- To convert from the mass of a compound to the number of molecules or formula units, use the molar mass of the compound to convert from grams to moles, then use Avogadro's number to convert from moles to number of molecules or formula units.

- A chemical formula gives equivalences between the number of atoms of each element in the compound and between the number of moles of each element in the compound.

- To convert from the mass of a compound to the mass of a constituent element, use the molar mass of the compound to convert from the mass of the compound to the number of moles of compound, use the chemical formula to convert from the number of moles of compound to the number of moles of the constituent element, and then use the molar mass of the element to convert from the number of moles of the element to the mass of the element.

- Mass percent composition, or simply mass percent, of an element in a compound is the element's percentage of the total mass of the compound. It can be computed from experimental data or from the chemical formula.

- Mass percent composition can be used as a conversion factor between the mass of a constituent element and the mass of the compound.

- An empirical formula gives the smallest whole number ratio of each type of atom in the molecule. A molecular formula gives the actual number of each type of atom in the molecule.

- Empirical formulas can be calculated from experimental data. Convert the given masses of each element to number of moles of each element. If the percent composition is given, assume a 100-g sample and compute the mass of each element from the given percentages. Write a pseudoformula for the compound using the number of moles of each element. Divide each subscript in the formula by the smallest subscript. If the subscripts are not whole numbers, multiply all the subscripts by a small whole number to obtain whole-number subscripts.

Chemical Composition

- A molecular formula is a whole-number multiple of the empirical formula of the compound. The molar mass of a compound is a whole-number multiple of the empirical formula mass.

- To calculate the molecular formula from the empirical formula and molar mass of a compound, first compute the empirical formula molar mass. Next, find n, the ratio of the molar mass to the empirical formula molar mass. Multiply the subscripts in the empirical formula by n to obtain the molecular formula.

SKILLBUILDER PROBLEMS AND SOLUTIONS

SKILLBUILDER 6.1 Converting between Moles and Numbers of Atoms

How many gold atoms are in a pure gold ring containing 8.83×10^{-2} mol Au?

Given: 8.83×10^{-2} mol Au

Find: Au atoms

Conversion Factor:

1 mol Au = 6.022×10^{23} Au atoms

Solution Map:

mol Au → Au atoms

$$\frac{6.022 \times 10^{23} \text{ Au atoms}}{1 \text{ mol Au}}$$

Solution:

$$8.83 \times 10^{-2} \text{ mol Au} \times \frac{6.022 \times 10^{23} \text{ Au atoms}}{1 \text{ mol Au}} = 5.32 \times 10^{22} \text{ Au atoms}$$

SKILLBUILDER 6.2 The Mole Concept—Converting between Grams and Moles

Calculate the number of grams of sulfur in 2.78 mol of sulfur.

Given: 2.78 mol S

Find: g S

Chapter 6

Conversion Factor: 32.07 g S = 1 mol S

Solution Map:

$$\boxed{\text{mol S}} \rightarrow \boxed{\text{g S}}$$

$$\frac{32.07 \text{ g S}}{1 \text{ mol S}}$$

Solution:

$$2.78 \text{ mol S} \times \frac{32.07 \text{ g S}}{1 \text{ mol S}} = 89.2 \text{ g S}$$

SKILLBUILDER 6.3   The Mole Concept—Converting Between Grams and Number of Atoms

Calculate the mass of 1.23×10^{24} helium atoms.

Given: 1.23×10^{24} He atoms

Find: g He

Conversion Factors:

1 mol He = 6.022×10^{23} He atoms

4.00 g He = 1 mol He

Solution Map:

$$\boxed{\text{He atoms}} \rightarrow \boxed{\text{mol He}} \rightarrow \boxed{\text{g He}}$$

$$\frac{1 \text{ mol He}}{6.022 \times 10^{23} \text{ He atoms}} \quad \frac{4.00 \text{ g He}}{1 \text{ mol He}}$$

Solution:

$$1.23 \times 10^{24} \text{ He atoms} \times \frac{1 \text{ mol He}}{6.022 \times 10^{23} \text{ Au atom}} \times \frac{4.00 \text{ g He}}{1 \text{ mol He}} = 8.17 \text{ g He}$$

SKILLBUILDER 6.4   The Mole Concept—Converting between Grams and Moles

Calculate the number of moles of NO_2 in 1.18 g of NO_2.

Chemical Composition

Given: 1.18 g NO$_2$

Find: mol NO$_2$

Conversion Factors:

NO$_2$ molar mass = 1(Molar mass of N) + 2(Molar mass of O)
   = 1(14.01 g/mol) + 2(16.00 g/mol)
   = 46.01 g/mol

Solution Map:

$$\boxed{\text{g NO}_2} \rightarrow \boxed{\text{mol NO}_2}$$

$$\frac{1 \text{ mol NO}_2}{46.01 \text{ g NO}_2}$$

Solution:

$$1.18 \text{ g NO}_2 \times \frac{1 \text{ mol NO}_2}{46.01 \text{ g NO}_2} = 0.0256 \text{ mol NO}_2$$

SKILLBUILDER 6.5   The Mole Concept—Converting between Mass and Number of Molecules

How many H$_2$O molecules are in a sample of water with a mass of 3.64 g?

Given: 3.64 g H$_2$O

Find: H$_2$O molecules

Conversion Factors:

1 mol H$_2$O = 6.022 × 10^{23} H$_2$O molecules

18.02 g H$_2$O = 1 mol H$_2$O

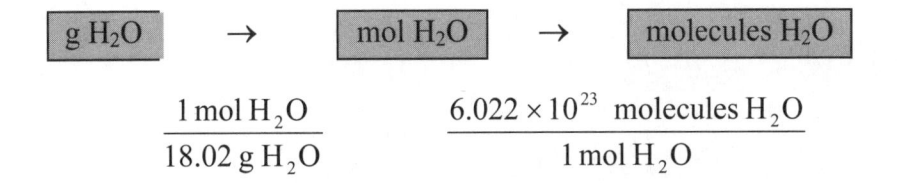

Solution Map:

$$\boxed{\text{g H}_2\text{O}} \rightarrow \boxed{\text{mol H}_2\text{O}} \rightarrow \boxed{\text{molecules H}_2\text{O}}$$

$$\frac{1 \text{ mol H}_2\text{O}}{18.02 \text{ g H}_2\text{O}} \qquad \frac{6.022 \times 10^{23} \text{ molecules H}_2\text{O}}{1 \text{ mol H}_2\text{O}}$$

Chapter 6

Solution:

$$3.64 \text{ g H}_2\text{O} \times \frac{1 \text{ mol H}_2\text{O}}{18.02 \text{ g H}_2\text{O}} \times \frac{6.022 \times 10^{23} \text{ molecules H}_2\text{O}}{1 \text{ mol H}_2\text{O}} = 1.22 \times 10^{23} \text{ H}_2\text{O molec}$$

SKILLBUILDER 6.6   **Chemical Formulas as Conversion Factors—Converting between Moles of a Compound and Moles of a Constituent Element**

Determine the number of moles of O in 1.4 mol of H_2SO_4.

Given: 1.4 mol H_2SO_4

Find: mol O

Conversion Factor:

$$4 \text{ mol O} \equiv 1 \text{ mol H}_2\text{SO}_4$$

Solution Map:

mol H_2SO_4 → mol O

$$\frac{4 \text{ mol O}}{1 \text{ mol H}_2\text{SO}_4}$$

Solution:

$$1.4 \text{ mol H}_2\text{SO}_4 \times \frac{4 \text{ mol O}}{1 \text{ mol H}_2\text{SO}} = 5.6 \text{ mol O}$$

SKILLBUILDER 6.7   **Chemical Formulas as Conversion Factors—Converting between Grams of a Compound and Grams of a Constituent Element**

Determine the mass of oxygen in a 5.8-g sample of sodium bicarbonate ($NaHCO_3$).

Given: 5.8 g $NaHCO_3$

Find: g O

Conversion Factors:

We need the molar mass of sodium carbonate.

Chemical Composition

Molar mass = molar mass Na + molar mass H + molar mass C + 3(molar mass O)
= 1(22.99 g/mol) + 1(1.01 g/mol) + 1(12.01 g/mol) + 3(16.00 g/mol)
= 22.99 g/mol + 1.01 g/mol + 12.01 g/mol + 48.00 g/mol
= 84.01 g/mol

3 mol O ≡ 1 mol $NaHCO_3$

1 mol O = 16.00 g O

Solution Map:

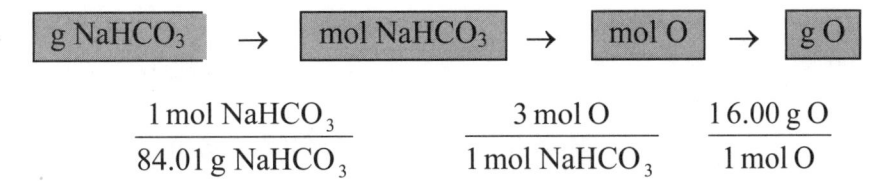

g $NaHCO_3$ → mol $NaHCO_3$ → mol O → g O

$$\frac{1 \text{ mol NaHCO}_3}{84.01 \text{ g NaHCO}_3} \qquad \frac{3 \text{ mol O}}{1 \text{ mol NaHCO}_3} \qquad \frac{16.00 \text{ g O}}{1 \text{ mol O}}$$

Solution:

$$5.8 \text{ g NaHCO}_3 \times \frac{1 \text{ mol NaHCO}_3}{84.01 \text{ g NaHCO}_3} \times \frac{3 \text{ mol O}}{1 \text{ mol NaHCO}_3} \times \frac{16.00 \text{ g O}}{1 \text{ mol O}} = 3.3 \text{ g O}$$

SKILLBUILDER PLUS

Determine the mass of oxygen in a 7.20-g sample of $Al_2(SO_4)_3$.

Given: 7.20 g $Al_2(SO_4)_3$

Find: g O

Conversion Factors:

We need the molar mass of aluminum sulfate.

Molar mass $Al_2(SO_4)_3$ = 2(molar mass Al) + 3(molar mass S) + 12(molar mass O)
= 2(26.98 g/mol) + 3(32.07 g/mol) + 12(16.00 g/mol)
= 53.96 + 96.21 + 192.00
= 342.17 g/mol

12 mol O ≡ 1 mol $Al_2(SO_4)_3$

1 mol O = 16.00 g O

Chapter 6

Solution Map:

$$\boxed{g\ Al_2(SO_4)_3} \rightarrow \boxed{mol\ Al_2(SO_3)_4} \rightarrow \boxed{mol\ O} \rightarrow \boxed{g\ O}$$

$$\frac{1\ mol\ Al_2(SO_4)_3}{342.17\ g\ Al_2(SO_4)_3} \qquad \frac{12\ mol\ O}{1\ mol\ Al_2(SO_4)_3} \qquad \frac{16.00\ g\ O}{1\ mol\ O}$$

Solution:

$$7.20\ g\ Al_2(SO_4)_3 \times \frac{1\ mol\ Al_2(SO_4)_3}{342.17\ g\ Al_2(SO_4)_3} \times \frac{12\ mol\ O}{1\ mol\ Al_2(SO_4)_3} \times \frac{16.00\ g\ O}{1\ mol\ O} = 4.04\ g\ O$$

SKILLBUILDER 6.8   **Using Mass Percent Composition as a Conversion Factor**

If someone consumes 22 g of sodium chloride, how much sodium does that person consume? Sodium chloride is 39% sodium by mass.

Given: 22 g NaCl

Find: g Na

Conversion Factor:

$$39\ g\ Na \equiv 100\ g\ NaCl$$

Solution Map:

$$\boxed{g\ NaCl} \rightarrow \boxed{g\ Na}$$

$$\frac{39\ g\ Na}{100\ g\ NaCl}$$

Solution:

$$22\ g\ NaCl \times \frac{39\ g\ Na}{100\ g\ NaCl} = 8.6\ g\ Na$$

SKILLBUILDER 6.9   **Mass Percent Composition**

Acetic acid ($HC_2H_3O_2$) is the active ingredient in vinegar. Calculate the mass percent composition of O in acetic acid.

Given: $HC_2H_3O_2$

Find: mass percent O

Chemical Composition

Equation:

$$\text{Mass percent of element X} = \frac{\text{Mass of element X in 1 mol of compound}}{\text{Mass of 1 mol of the compound}} \times 100\%$$

Conversion Factors:

We need the molar mass of $HC_2H_3O_2$.

Molar mass $HC_2H_3O_2$ = 4(molar mass H) + 2(molar mass C) + 2(molar mass O)
= 4(1.01 g/mol) + 2(12.01 g/mol) + 2(16.00 g/mol)
= 60.06 g/mol

2 mol O ≡ 1 mol $HC_2H_3O_2$

1 mol O = 16.00 g O

To solve this problem, we must find the two quantities required for the mass fraction equation. The first is the mass of O in 1 mol of $HC_2H_3O_2$.

Solution Map:

$$\boxed{\text{mol } HC_2H_3O_2} \rightarrow \boxed{\text{mol O}} \rightarrow \boxed{\text{g O}}$$

$$\frac{2 \text{ mol O}}{1 \text{ mol } HC_2H_3O_2} \qquad \frac{16.00 \text{ g O}}{1 \text{ mol O}}$$

Solution:

$$1 \text{ mol } HC_2H_3O_2 \times \frac{2 \text{ mol O}}{1 \text{ mol } HC_2H_3O_2} \times \frac{16.00 \text{ g O}}{1 \text{ mol O}} = 32.00 \text{ g O}$$

The second quantity is the mass of 1 mol of $HC_2H_3O_2$, which is simply its molar mass, 60.06 g. We now substitute these two quantities into the mass percent equation.

$$\text{Mass percent of O} = \frac{\text{Mass of element O in 1 mol of } HC_2H_3O_2}{\text{Mass of 1 mol of } HC_2H_3O_2} \times 100\%$$

$$= \frac{32.00 \text{ g O}}{60.06 \text{ g } HC_2H_3O_2} \times 100\%$$

$$= 53.28\%$$

Chapter 6

SKILLBUILDER 6.10   Obtaining an Empirical Formula from Experimental Data

A sample of a compound is decomposed in the laboratory and produces 165 g of carbon, 27.8 g of hydrogen, and 220.2 g O. Calculate the empirical formula of the compound.

Given: 165 g C
27.8 g H
220.2 g O

Find: empirical formula

Solution:

$$165 \text{ g C} \times \frac{1 \text{ mol C}}{12.01 \text{ g C}} = 13.7 \text{ mol C}$$

$$27.8 \text{ g H} \times \frac{1 \text{ mol H}}{1.01 \text{ g H}} = 27.5 \text{ mol H}$$

$$220.2 \text{ g O} \times \frac{1 \text{ mol O}}{16.00 \text{ g O}} = 13.76 \text{ mol O}$$

$C_{13.7}H_{27.5}O_{13.76}$

$C_{\frac{13.7}{13.7}}H_{\frac{27.5}{13.7}}O_{\frac{13.76}{13.7}}$

$C_{\frac{13.7}{13.7}}H_{\frac{27.5}{13.7}}O_{\frac{13.76}{13.7}} \rightarrow C_1H_2O_1$

The empirical formula is CH_2O.

SKILLBUILDER 6.11   Obtaining an Empirical Formula from Experimental Data

Ibuprofen, an aspirin substitute, has the following mass percent composition.

C 75.69%
H 8.80%
O 15.51%

Calculate the empirical formula of the compound.

Given: In a 100-g sample:

75.69 g C
8.80 g H
15.51 g O

Find: empirical formula

Solution:

$$75.69 \text{ g C} \times \frac{1 \text{ mol C}}{12.01 \text{ g C}} = 6.302 \text{ mol C}$$

$$8.80 \text{ g H} \times \frac{1 \text{ mol H}}{1.01 \text{ g H}} = 8.71 \text{ mol H}$$

$$15.51 \text{ g O} \times \frac{1 \text{ mol O}}{16.00 \text{ g O}} = 0.9694 \text{ mol O}$$

$C_{6.302}H_{8.71}O_{0.9694}$

$C_{\frac{6.302}{0.9694}}H_{\frac{8.71}{0.9694}}O_{\frac{0.9694}{0.9694}}$

$C_{\frac{6.302}{0.9694}}H_{\frac{8.71}{0.9694}}O_{\frac{0.9694}{0.9694}} \rightarrow C_{6.5}H_{9.0}O_1$

$C_{6.5}H_{9.0}O_1 \times 2 \rightarrow C_{13}H_{18}O_2$

The empirical formula is $C_{13}H_{18}O_2$.

SKILLBUILDER 6.12 — Calculating an Empirical Formula from Reaction Data

A 1.56-g sample of copper reacts with oxygen to form 1.95 g of the metal oxide. What is the formula of the oxide?

Given: 1.56 g Cu, 1.95 g oxide

Find: empirical formula

Solution:
You must recognize this problem as one requiring a special procedure and you must follow that procedure.

Chapter 6

1) Write down (or compute) the masses of each element present in the sample of the compound. In this case, we are given the mass of the initial Cu sample and the mass of its oxide after the sample reacts with oxygen. The oxide contains all of the copper originally present.

   1.56 g Cu

   The mass of oxygen is the difference between the mass of the oxide and the mass of copper.

   Mass O = Mass oxide − Mass copper
   = 1.95 g − 1.56 g
   = 0.39 g O

2) Convert each of the masses in step 1 to moles using the appropriate molar mass for each element as a conversion factor.

   $$1.56 \text{ g Cu} \times \frac{1 \text{ mol Cu}}{63.55 \text{ g C}} = 0.0245 \text{ mol Cu}$$

   $$0.39 \text{ g O} \times \frac{1 \text{ mol O}}{16.00 \text{ g H}} = 0.024 \text{ mol O}$$

3) Write down a pseudoformula for the compound using the moles of each element obtained in step 2 as subscripts.

   $Cu_{0.0245}O_{0.024}$

4) Divide all the subscripts in the formula by the smallest subscript.

   $$Cu_{\frac{0.0245}{0.024}}O_{\frac{0.024}{0.024}} \rightarrow CuO$$

The empirical formula is CuO.

SKILLBUILDER 6.13 Calculating Molecular Formula from Empirical Formula and Molar Mass

Butane is a compound containing carbon and hydrogen that is used as a fuel in butane lighters. Its empirical formula is C_2H_5 and its molar mass is 58.12 g/mol. Find its molecular formula.

Given: empirical formula = C_2H_5; molar mass = 58.12 g/mol

Find: molecular formula

Solution:

Chemical Composition

The molecular formula is *n* times the empirical formula. To find *n*, divide the molar mass by the empirical formula molar mass.

$$\text{Empirical formula molar mass} = 2(\text{molar mass C}) + 5(\text{molar mass H})$$
$$= 2(12.01 \text{ g/mol}) + 5(1.01 \text{ g/mol})$$
$$= 29.07 \text{ g/mol}$$

$$n = \frac{\text{Molar mass}}{\text{Empirical formula mass}} = \frac{58.12 \text{ g/mol}}{29.07 \text{ g/mol}} = 2$$

Therefore, the molecular formula is 2 times the empirical formula.

$$\text{Molecular formula} = C_2H_5 \times 2 = C_4H_{10}$$

SKILLBUILDER PLUS

A compound with the following mass percent composition has a molar mass of 60.10 g/mol. Find its molecular formula.

C: 39.97%   H: 13.41%   N: 46.62%

Given: molar mass = 60.10 g/mol

In a 100-g sample,

39.97 g C
13.41 g H
46.62 g N

Find: molecular formula

Solution:
We must first find the empirical formula.

$$39.97 \text{ g C} \times \frac{1 \text{ mol C}}{12.01 \text{ g C}} = 3.328 \text{ mol C}$$

$$13.41 \text{ g H} \times \frac{1 \text{ mol H}}{1.01 \text{ g H}} = 13.3 \text{ mol H}$$

$$46.62 \text{ g N} \times \frac{1 \text{ mol N}}{14.01 \text{ g N}} = 3.328 \text{ mol N}$$

$C_{3.328}H_{13.3}N_{3.328}$

Chapter 6

$$C_{\frac{3.328}{3.328}} H_{\frac{13.3}{3.328}} N_{\frac{3.328}{3.328}}$$

$$C_{\frac{3.328}{3.328}} H_{\frac{13.3}{3.328}} N_{\frac{3.328}{3.328}} \rightarrow CH_4N$$

The empirical formula is CH₄N.

The molecular formula is *n* times the empirical formula. To find *n*, we divide the molar mass by the empirical formula molar mass.

Empirical formula molar mass = molar mass C + 4(molar mass H) + molar mass N
= 12.01 g/mol + 4(1.01 g/mol) + 14.01 g/mol
= 30.06 g/mol

$$n = \frac{\text{Molar mass}}{\text{Empirical formula mass}} = \frac{60.10 \text{ g/mol}}{30.06 \text{ g/mol}} = 2$$

Therefore, the molecular formula is 2 times the empirical formula.

Molecular formula = CH₄N × 2 = C₂H₈N₂

SELF-TEST QUESTIONS

A. Match the following terms with the phrases below.

> empirical formula   molar mass
> empirical formula molar mass   mole (mol)
> mass percent (composition)   molecular formula

1. Quantity used by chemists with a value of 6.022×10^{23}
2. Mass of one mole of atoms of an element or one mole of molecules or formula units of a compound
3. An element's percentage of the total mass of a compound
4. Formula that gives the specific number of each type of atom in a molecule
5. Formula that gives the smallest whole number ratio of each type of atom in a molecule
6. Sum of the masses of all the atoms in the empirical formula

B. True/False

1. Avogadro's number is 6.022×10^{23}.
2. One mole of He atoms weighs 4.00 amu.
3. The mass of 1 mole of atoms of an element is its molar mass.
4. One mole of sodium atoms weighs 32.07 g.

Chemical Composition

5. One mole of oxygen molecules weighs 16.00 g.
6. One mole of water weighs 18.02 g.
7. There are 2 moles of Cl atoms for every 1 mole of CH_2Cl_2.
8. The mass percent of C in CO is 50%.
9. The empirical formula for hydrogen peroxide, H_2O_2, is HO.
10. The molecular formula of a compound can be determined from its empirical formula and its molar mass.

C. Multiple Choice

1. How many silver atoms are in the 0.40 mol Ag?
    a) 2.4×10^{23}
    b) 6.0×10^{23}
    c) 1.5×10^{24}
    d) 2.4×10^{24}

2. How many moles of gold atoms are in 1.63×10^{22} atoms of gold?
    a) 0.0271 mol
    b) 4.39 mol
    c) 5.86 mol
    d) 36.9 mol

3. What is the mass, in grams, of 0.522 mol copper?
    a) 0.522 g
    b) 33.2 g
    c) 63.6 g
    d) 3.14×10^{23} g

4. How many sugar molecules, $C_{12}H_{22}O_{11}$, are in 30.0 g of sugar?
    a) 5.28×10^{22} molecules
    b) 2.01×10^{23} molecules
    c) 1.81×10^{25} molecules
    d) 6.18×10^{27} molecules

5. What is the mass, in grams, of 3.18×10^{21} formula units of $Ca_3(PO_4)_2$?
    a) 0.882 g
    b) 1.31 g
    c) 1.47 g
    d) 1.64 g

6. How many moles of H are in 0.424 mol $Ca(C_2H_3O_2)_2$?
    a) 0.848 mol H
    b) 1.27 mol H
    c) 2.12 mol H
    d) 2.54 mol H

Chapter 6

7. Calculate the mass percent S in Na$_2$S$_2$O$_3$.
   a) 20.28%
   b) 31.13%
   c) 40.56%
   d) 45.13%

8. A compound consists of 25.94% N and 74.06% O by mass. Determine the empirical formula of the compound.
   a) NO
   b) NO$_3$
   c) N$_2$O$_4$
   d) N$_2$O$_5$

9. A compound is formed by the combination of 13.99 g iron with 6.01 g oxygen. What is the empirical formula of the compound?
   a) FeO
   b) Fe$_2$O
   c) FeO$_2$
   d) Fe$_2$O$_3$

10. A compound has an empirical formula of CH$_2$ and a molar mass of 84.18 g/mol. Find its molecular formula.
    a) C$_4$H$_8$
    b) C$_5$H$_{10}$
    c) C$_6$H$_{12}$
    d) C$_7$H$_{14}$

D. Crossword Puzzle

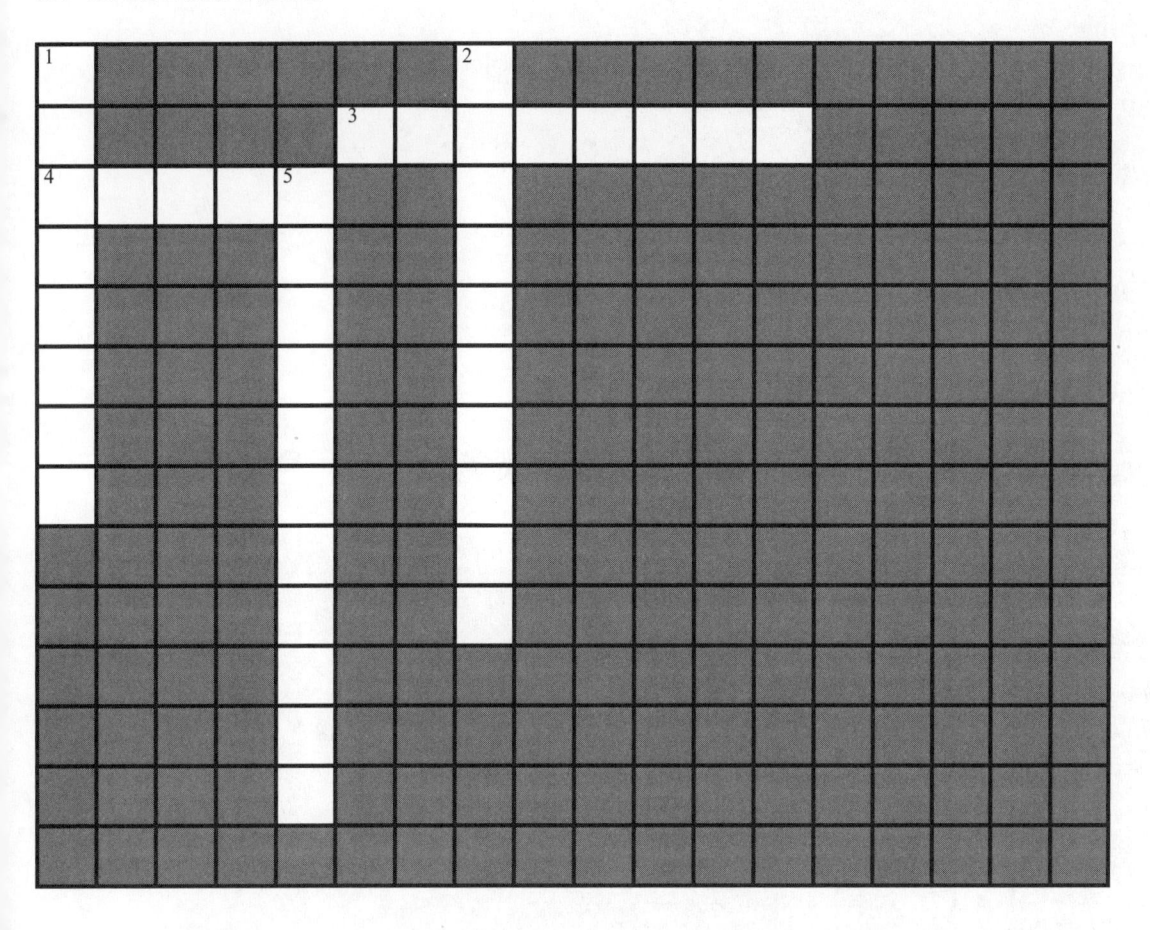

ACROSS

3. Ion added to drinking water to prevent tooth decay
4. Molecule in the upper atmosphere with chemical formula O_3

DOWN

1. Chemist whose name is associated with the mole
2. Numbers in a chemical formula
5. Relationship represented by "≡" sign

Chapter 6

ANSWERS TO SELF-TEST QUESTIONS

A. Matching
1. mole 2. molar mass 3. mass percent (composition) 4. molecular formula 5. empirical formula 6. empirical formula molar mass

B. True/False
1. T 2. F 3. T 4. F 5. F 6. T 7. T 8. F 9. T 10. T

C. Multiple Choice
1. a 2. a 3. b 4. a 5. d 6. d 7. c 8. d 9. d 10. c

D. Crossword Puzzle

	1						2					
	A						S					
	V				3F	L	U	O	R	I	D	E
	4O	Z	O	N	5E		B					
	G				Q		S					
	A				U		C					
	D				I		R					
	R				V		I					
	O				A		P					
					L		T					
					E		S					
					N							
					C							
					E							

84

Chemical Reactions

CHAPTER OVERVIEW

The topic of Chapter 7 is chemical reactions. Examples of evidence for chemical reactions are described. Precipitation, acid-base, gas evolution, redox, and combustion reactions are discussed.

CHAPTER OBJECTIVES

After reading and studying the text, students should be able to:

1. Give examples of observations that provide evidence for chemical reactions.
2. Identify reactants and products in a chemical equation.
3. List abbreviations that indicate the states of reactants and products in chemical equations.
4. Write balanced chemical equations, given the names of reactants and products.
5. Use solubility rules to determine if a compound is soluble or insoluble in water.
6. Write equations for precipitation reactions.
7. Write molecular, complete ionic, and net ionic equations for reactions in solution.
8. Write balanced equations for acid-base reactions.
9. Write balanced equations for gas evolution reactions.
10. Identify redox reactions.
11. Write balanced equations for combustion reactions.
12. Classify chemical reactions as precipitation reactions, acid-base reactions, gas evolution reactions, oxidation-reduction, or combustion reactions.
13. Classify reactions as synthesis reactions, decomposition reactions, single displacement reactions, or double displacement reactions.

CHAPTER IN REVIEW

- A chemical reaction is a change of one or more substances into different substances.

- Gas evolution reactions are reactions that occur in liquids and form a gas.

- Reactions in which a substance reacts with oxygen, emits heat, and forms one or more oxygen containing compounds are called combustion reactions.

- Oxidation-reduction reactions are reactions in which electrons are transferred from one substance to another.

Chapter 7

- Reactions that form solid substances in water are called precipitation reactions.

- Evidence for chemical reactions includes color changes, formation of a solid in a previously clear solution, formation of a gas after adding a substance to a solution, heat absorption or emission, and light emission. Such changes are not definitive evidence; only chemical analysis showing that the initial substances have changed into other substances conclusively proves that a chemical reaction has occurred.

- Chemical reactions are represented with chemical equations. The substances on the left side of the equation are called reactants; the substances on the right side of the equation are called products. Physical states of reactants and products are often specified in parentheses next to the chemical formulas.

- Atoms are neither created nor destroyed in a chemical reaction.

- Chemical equations are balanced by adding coefficients to ensure that the number of each type of atom on both sides of the equation is equal.

- To balance chemical equations, first write a skeletal equation by writing chemical formulas for reactants and products. Balance any metal that occurs in only one compound on both sides of the equation, then balance any nonmetal that occurs in only one compound on both sides of the equation. If an element occurs as a free element on either side of the chemical equation, balance that element last. If the balanced equation contains coefficient fractions, clear these by multiplying the entire equation by the appropriate factor to give whole-number coefficients. Check that the equation is balanced by summing the total number of each type of atom on both sides of the equation.

- An aqueous solution is a mixture of a substance with water.

- Strong electrolyte solutions are solutions that contain dissolved ionic compounds.

- A compound is soluble if it dissolves in water; it is insoluble if it does not dissolve in water.

- Solubility rules are empirical rules that have been deduced from observations of the solubility of many compounds.

- Precipitation reactions occur when two aqueous solutions are mixed and an insoluble compound forms.

- To write equations for precipitation reactions, write the formulas of the two compounds being mixed as reactants in a chemical equation. Below the chemical equation, write the formulas of the potentially insoluble products that could form from the reactants by combining the cations of each reactant with the anion from the other reactant. Use the solubility rules to determine if any formulas represent insoluble compounds. If all of the formulas represent soluble compounds, no precipitate forms. If no precipitate forms, write *NO REACTION* on the right side of the arrow. If one or both formulas represent insoluble compounds, write their formulas on the product side of the equation. Use (*s*) to indicate a solid, use (*aq*) to indicate aqueous. Balance the equation.

- Molecular equations show complete neutral formulas for every compound in the reaction. Complete ionic equations show the reactants and products as they are actually present in solution.

- Spectator ions appear unchanged on both sides of a complete ionic equation. Spectator ions do not participate in the reaction.

- Net ionic equations show only the species that actually change during a reaction.

- Acid-base reactions generally form water and a salt.

- The net ionic equation for many acid-base reactions is
  $H^+(aq) + OH^-(aq) \rightarrow H_2O(l)$.

- The main types of compounds that can form gases in aqueous solution include sulfides, carbonates and bicarbonates, sulfites and bisulfites, and compounds containing the ammonium ion.

- Oxidation is loss of electrons; reduction is gain of electrons.

- Oxidation-reduction reactions are often called redox reactions. They are reactions in which one substance transfers electrons to another substance. Reactions of metals with nonmetals are redox reactions. Combustion reactions are redox reactions in which a substance reacts with elemental oxygen to form one or more oxygen-containing compounds.

- Synthesis or combination reactions are reactions in which simpler substances combine to form more complex substances. Decomposition reactions are reactions in which a complex substance breaks apart to form simpler substances.

- Displacement or single displacement reactions are reactions in which one element replaces another in a compound. Double displacement reactions are reactions in which two elements or groups of elements in two different compounds exchange places to form two different compounds.

Chapter 7

SKILLBUILDER PROBLEMS AND SOLUTIONS

SKILLBUILDER 7.1   Evidence of a Chemical Reaction

Which of the following is a chemical reaction? Why?

a) butane burning in a butane lighter
b) butane evaporating out of a butane lighter
c) wood burning
d) dry ice subliming

Solution:
a) Chemical reaction; butane reacts with oxygen, heat and light are emitted.
b) Not a chemical reaction; liquid and gaseous butane are both butane.
c) Chemical reaction; wood reacts with oxygen, heat and light are emitted
d) Not a chemical reaction; dry ice is solid carbon dioxide which sublimes to gaseous carbon dioxide.

SKILLBUILDER 7.2   Writing Balanced Chemical Equations

Write a balanced equation for the reaction between solid chromium(III) oxide and solid carbon to produce solid chromium and carbon dioxide gas.

Solution:

$$Cr_2O_3(s) + C(s) \rightarrow Cr(s) + CO_2(g)$$

Begin with O:

$$Cr_2O_3(s) + C(s) \rightarrow Cr(s) + CO_2(g)$$

3 O atoms → 2 O atoms

To balance O, put a 2 in front of $Cr_2O_3(s)$ and a 3 in front of $CO_2(g)$.

$$2\ Cr_2O_3(s) + C(s) \rightarrow Cr(s) + 3\ CO_2(g)$$

6 O atoms → 6 O atoms

Balance Cr:

$$2\ Cr_2O_3(s) + C(s) \rightarrow Cr(s) + 3\ CO_2(g)$$

4 Cr atoms → 1 Cr atom

To balance Cr, put a 4 before Cr(s).

$$2\ Cr_2O_3(s) + C(s) \rightarrow \mathbf{4}\ Cr(s) + 3\ CO_2(g)$$

4 Cr atoms → 4 Cr atoms

Balance C:

$$2\ Cr_2O_3(s) + C(s) \rightarrow 4\ Cr(s) + 3\ CO_2(g)$$

1 C atom → 3 C atoms

To balance C, put a 3 before C(s).

$$2\ Cr_2O_3(s) + \mathbf{3}\ C(s) \rightarrow 4\ Cr(s) + 3\ CO_2(g)$$

3 C atoms → 3 C atoms

$$2\ Cr_2O_3(s) + 3\ C(s) \rightarrow 4\ Cr(s) + 3\ CO_2(g)$$

Reactants		Products
4 Cr atoms	→	4 Cr atoms
6 O atoms	→	6 O atoms
3 C atoms	→	3 C atoms

The equation is balanced.

SKILLBUILDER 7.3   **Writing Balanced Chemical Equations**

Write a balanced equation for the combustion of gaseous C_4H_{10} in which it combines with gaseous oxygen to form gaseous carbon dioxide and gaseous water.

Solution:

$$C_4H_{10}(g) + O_2(g) \rightarrow CO_2(g) + H_2O(g)$$

Begin with C:

$$C_4H_{10}(g) + O_2(g) \rightarrow CO_2(g) + H_2O(g)$$

4 C atoms → 1 C atom

To balance C, put a 4 in front of $CO_2(g)$.

Chapter 7

$$C_4H_{10}(g) + O_2(g) \rightarrow 4\,CO_2(g) + H_2O(g)$$

4 C atoms → 4 C atoms

Balance H next:

$$C_4H_{10}(g) + O_2(g) \rightarrow 4\,CO_2(g) + H_2O(g)$$

10 H atoms → 2 H atoms

To balance H, put a 5 before $H_2O(g)$.

$$C_4H_{10}(g) + O_2(g) \rightarrow 4\,CO_2(g) + 5\,H_2O(g)$$

10 H atoms → 10 H atoms

Balance O:

$$C_4H_{10}(g) + O_2(g) \rightarrow 4\,CO_2(g) + 5\,H_2O(g)$$

2 O atoms → 8 O atoms + 5 O atoms = 13 O atoms

To balance O, put a $\frac{13}{2}$ before $O_2(g)$.

$$C_4H_{10}(g) + \tfrac{13}{2}O_2(g) \rightarrow 4\,CO_2(g) + 5\,H_2O(g)$$

13 O atoms → 8 O atoms + 5 O atoms = 13 O atoms

Clear the coefficient fraction by multiplying by 2.

$$[C_4H_{10}(g) + \tfrac{13}{2}O_2(g) \rightarrow 4\,CO_2(g) + 5\,H_2O(g)] \times 2$$

$$2\,C_4H_{10}(g) + 13\,O_2(g) \rightarrow 8\,CO_2(g) + 10\,H_2O(g)$$

$$2\,C_4H_{10}(g) + 13\,O_2(g) \rightarrow 8\,CO_2(g) + 10\,H_2O(g)$$

<u>Reactants</u>		<u>Products</u>
8 C atoms	→	8 C atoms
20 H atoms	→	20 H atoms
26 O atoms	→	26 O atoms

Chemical Reactions

> **SKILLBUILDER 7.4**   **Balancing Chemical Equations**

Write a balanced equation for the reaction of aqueous lead(II) acetate with aqueous potassium iodide to form solid lead(II) iodide and aqueous potassium acetate.

Solution:
We first write a skeletal equation containing formulas for each of the reactants and products.

$$Pb(C_2H_3O_2)_2(aq) + KI(aq) \rightarrow PbI_2(s) + KC_2H_3O_2(aq)$$

The formulas for each compound MUST BE CORRECT before we begin to balance the equation. Carbon, hydrogen, and oxygen are in a polyatomic ion that stays intact on both sides of the equation, so we will balance these elements as a group. There are 2 $C_2H_3O_2^-$ ions on the left hand side of the equation, so we need 2 on the right. Place a 2 in front of $KC_2H_3O_2(aq)$ to balance the $C_2H_3O_2^-$ ion.

$$Pb(C_2H_3O_2)_2(aq) + KI(aq) \rightarrow PbI_2(s) + \mathbf{2}\ KC_2H_3O_2(aq)$$

We balance K next. Since there are 2 K atoms on the right side of the equation, we place a 2 in front of KI(aq) on the left side of the equation.

$$Pb(C_2H_3O_2)_2(aq) + \mathbf{2}\ KI(aq) \rightarrow PbI_2(s) + 2\ KC_2H_3O_2(aq)$$

We balance Pb next. There is 1 Pb atom on each side of the equation. Pb is balanced.

$$Pb(C_2H_3O_2)_2(aq) + \mathbf{2}\ KI(aq) \rightarrow PbI_2(s) + 2\ KC_2H_3O_2(aq)$$

Next we balance I. There are 2 I atoms on each side of the equation. I is balanced.

$$Pb(C_2H_3O_2)_2(aq) + \mathbf{2}\ KI(aq) \rightarrow PbI_2(s) + 2\ KC_2H_3O_2(aq)$$

Reactants		Products
1 Pb atom	→	1 Pb atom
4 C atoms	→	4 O atoms
6 H atoms	→	6 H atoms
4 O atoms	→	4 O atoms
2 K atoms	→	2 K atoms
2 I atoms	→	2 I atoms

The equation is balanced.

Chapter 7

SKILLBUILDER 7.5   Balancing Chemical Equations

Balance the following chemical equation.

$$HCl(g) + O_2(g) \rightarrow H_2O(l) + Cl_2(g)$$

Solution:
Since H occurs in only one compound on each side of the equation, we balance it first. There is 1 H atom on the left-hand side of the equation and 2 H atoms on the right-hand side. We balance H by placing a 2 in front of HCl.

$$2\,HCl(g) + O_2(g) \rightarrow H_2O(l) + Cl_2(g)$$

Since O and Cl occur as free elements, we balance them last. There are 2 Cl atoms on the left side of the equation and 2 Cl atoms on the right, so Cl is balanced. There are 2 O atoms on the left and 1 O atom on the right. We balance O by adjusting the coefficient on O_2 (that way we don't alter other elements that are already balanced). Balance O by placing a $\frac{1}{2}$ in front of O_2.

$$2\,HCl(g) + \tfrac{1}{2}O_2(g) \rightarrow H_2O(l) + Cl_2(g)$$

Since the equation contains a coefficient fraction, we clear it by multiplying the entire equation by 2.

$$[2\,HCl(g) + \tfrac{1}{2}O_2(g) \rightarrow H_2O(l) + Cl_2(g)] \times 2$$

$$4\,HCl(g) + O_2(g) \rightarrow 2\,H_2O(l) + 2\,Cl_2(g)$$

Finally, we sum the number of each type of atom on each side to check that the equation is balanced.

$$4\,HCl(g) + O_2(g) \rightarrow 2\,H_2O(l) + 2\,Cl_2(g)$$

Reactants		Products
4 H atoms	→	4 H atoms
4 Cl atoms	→	4 Cl atoms
2 O atoms	→	2 O atoms

The equation is balanced.

Chemical Reactions

SKILLBUILDER 7.6   **Determining Whether a Compound Is Soluble**

Determine whether each of the following compounds is soluble or insoluble.

a) CuS
b) FeSO$_4$
c) PbCO$_3$
d) NH$_4$Cl

Solution:

a) Insoluble; compounds containing S^{2-} are insoluble, except when S^{2-} is paired with Li$^+$, Na$^+$, K$^+$, NH$_4^+$, Ca^{2+}, Sr^{2+}, or Ba^{2+}.
b) Soluble; compounds containing SO$_4^{2-}$ are soluble, except when SO$_4^{2-}$ is paired with Ca^{2+}, Sr^{2+}, Ba^{2+}, or Pb^{2+}.
c) Insoluble; compounds containing CO$_3^{2-}$ are insoluble, except when CO$_3^{2-}$ is paired with Li$^+$, Na$^+$, K$^+$, or NH$_4^+$.
d) Soluble; Compounds containing NH$_4^+$ are soluble.

SKILLBUILDER 7.7   **Writing Equations for Precipitation Reactions**

Write an equation for the precipitation reaction that occurs (if any) when solutions of potassium hydroxide and nickel(II) bromide are mixed.

Solution:

$$KOH(aq) + NiBr_2(aq) \rightarrow$$

Potentially Insoluble Products

    KBr   Ni(OH)$_2$

KrBr is *soluble* (compounds containing Br$^-$ are usually soluble and K$^+$ is not an exception).

Ni(OH)$_2$ is *insoluble* (compounds containing OH$^-$ are usually insoluble and Ni^{2+} is not an exception).

$$KOH(aq) + NiBr_2(aq) \rightarrow KBr(aq) + Ni(OH)_2(s)$$

Balance the equation.

$$2\ KOH(aq) + NiBr_2(aq) \rightarrow 2\ KBr(aq) + Ni(OH)_2(s)$$

Chapter 7

SKILLBUILDER 7.8 — Writing Equations for Precipitation Reactions

Write an equation for the precipitation reaction that occurs (if any) when solutions of ammonium chloride and iron(III) nitrate are mixed.

Solution:

$$NH_4Cl(aq) + Fe(NO_3)_3(aq) \rightarrow$$

Potentially Insoluble Products

NH_4NO_3 $FeCl_3$

NH_4NO_3 is *soluble* (compounds containing NH_4^+ are soluble).

$FeCl_3$ is *soluble* (compounds containing Cl^- are usually soluble and Fe^{3+} is not an exception).

$$NH_4Cl(aq) + Fe(NO_3)_3(aq) \rightarrow \text{NO REACTION}$$

SKILLBUILDER 7.9 — Predicting and Writing Equations for Precipitation Reactions

Write an equation for the precipitation reaction that occurs (if any) when solutions of potassium sulfate and strontium nitrate are mixed. If no reaction occurs, write *NO REACTION*.

Solution:

$$K_2SO_4(aq) + Sr(NO_3)_2(aq) \rightarrow$$

Potentially Insoluble Products

KNO_3 $SrSO_4$

KrBr is *soluble* (compounds containing K^+ are soluble).

$SrSO_4$ is *insoluble* (compounds containing SO_4^{2-} are normally soluble but Sr^{2+} is an exception.

$$K_2SO_4(aq) + Sr(NO_3)_2(aq) \rightarrow KNO_3(aq) + SrSO_4(s)$$

Balance the equation.

$$K_2SO_4(aq) + Sr(NO_3)_2(aq) \rightarrow 2\ KNO_3(aq) + SrSO_4(s)$$

Chemical Reactions

SKILLBUILDER 7.10   Writing Complete Ionic and Net Ionic Equations

Consider the following reaction occurring in aqueous solution.

$$2\ HBr(aq) + Ca(OH)_2(aq) \rightarrow 2\ H_2O(l) + CaBr_2(aq)$$

Write a complete ionic equation and net ionic equation for this reaction.

Solution:
We write the complete ionic equation by separating aqueous ionic compounds into their constituent ions. The $H_2O(l)$ remains as one unit.

Complete ionic equation:

$$2\ H^+(aq) + 2\ Br^-(aq) + Ca^{2+}(aq) + 2\ OH^-(aq) \rightarrow$$
$$2\ H_2O(l) + Ca^{2+}(aq) + 2\ Br^-(aq)$$

The net ionic equation eliminates the spectator ions, those that are not changing during the reaction.

Net ionic equation:

$$2\ H^+(aq) + 2\ OH^-(aq) \rightarrow 2\ H_2O(l)$$

or simply

$$H^+(aq) + OH^-(aq) \rightarrow H_2O(l)$$

SKILLBUILDER 7.11   Writing Equations for Acid-Base Reactions

Write a molecular and a net ionic equation for the reaction that occurs between aqueous H_2SO_4 and aqueous KOH.

Solution:
We must recognize these substances as an acid and a base. Acid-base reactions generally form water and an ionic compound (or salt) that usually remains dissolved in solution. We first write the skeletal reaction following this general pattern.

Molecular equation:

$$\underset{\text{acid}}{H_2SO_4(aq)} + \underset{\text{base}}{KOH(aq)} \rightarrow \underset{\text{water}}{H_2O(l)} + \underset{\text{salt}}{K_2SO_4(aq)}$$

We then balance the equation.

$$H_2SO_4(aq) + 2\ KOH(aq) \rightarrow 2\ H_2O(l) + K_2SO_4(aq)$$

Chapter 7

Complete ionic equation:

$$2\,H^+(aq) + SO_4^{2-}(aq) + 2\,K^+(aq) + 2\,OH^-(aq) \rightarrow 2\,H_2O(l) + 2\,K^+(aq) + SO_4^{2-}(aq)$$

The net ionic equation eliminates the spectator ions, those that are not changing during the reaction.

Net ionic equation:

$$2\,H^+(aq) + 2\,OH^-(aq) \rightarrow 2\,H_2O(l)$$

or simply

$$H^+(aq) + OH^-(aq) \rightarrow H_2O(l)$$

SKILLBUILDER 7.12   Writing Equations for Gas Evolution Reactions

Write a molecular equation for the gas evolution reaction that occurs when you mix aqueous hydrobromic acid and aqueous potassium sulfite.

Solution:
We begin by writing a skeletal equation that includes the reactants and products that form when the cation of each reactant reacts with the anion of the other.

$$HBr(aq) + K_2SO_3(aq) \rightarrow KBr(aq) + H_2SO_3(aq)$$

We recognize that $H_2SO_3(aq)$ decomposes into $H_2O(l)$ and $SO_2(g)$ and write the corresponding equation.

$$HBr(aq) + K_2SO_3(aq) \rightarrow KBr(aq) + H_2O(l) + SO_2(g)$$

Finally, we balance the equation.

$$2\,HBr(aq) + K_2SO_3(aq) \rightarrow 2\,KBr(aq) + H_2O(l) + SO_2(g)$$

SKILLBUILDER PLUS

Write a net ionic equation for the previous reaction.

Complete ionic equation:

$$2\,H^+(aq) + 2\,Br^-(aq) + 2\,K^+(aq) + SO_3^{2-}(aq) \rightarrow 2\,K^+(aq) + 2\,Br^-(aq) + H_2O(l) + SO_2(g)$$

Chemical Reactions

The net ionic equation eliminates the spectator ions, those that are not changing during the reaction.

Net ionic equation:

$$2\,H^+(aq) + SO_3^{2-}(aq) \rightarrow H_2O(l) + SO_2(g)$$

SKILLBUILDER 7.13   Identifying Redox Reactions

Which of the following is a redox reaction?

a) $2\,Li(s) + Cl_2(g) \rightarrow 2\,LiCl(s)$
b) $2\,Al(s) + 3\,Sn^{2+}(aq) \rightarrow 2\,Al^{3+}(aq) + 3\,Sn(s)$
c) $Pb(NO_3)_2(aq) + 2\,LiCl(aq) \rightarrow PbCl_2(s) + 2\,LiNO_3(aq)$
d) $C(s) + O_2(g) \rightarrow CO_2(g)$

Solution:
a) Redox reaction; lithium metal reacts with chlorine, a nonmetal.
b) Redox reaction; Al transfers electrons to Sn^{2+}.
c) Not a redox reaction; this is a precipitation reaction.
d) Redox reaction; carbon reacts with elemental oxygen.

SKILLBUILDER 7.14   Writing Combustion Reactions

Write a balanced equation for the combustion of liquid pentane (C_5H_{12}), a component of gasoline.

Solution:
Begin by writing a skeletal equation showing the reaction of C_5H_{12} with O_2 to form CO_2 and H_2O.

$$C_5H_{12}(l) + O_2(g) \rightarrow CO_2(g) + H_2O(g)$$

Balance the skeletal equation.

$$C_5H_{12}(l) + 8\,O_2(g) \rightarrow 5\,CO_2(g) + 6\,H_2O(g)$$

SKILLBUILDER PLUS

Write a balanced equation for the combustion of liquid ethanol (C_2H_5OH).

Chapter 7

Solution:
Begin by writing a skeletal equation showing the reaction of C_2H_5OH with O_2 to form CO_2 and H_2O.

$$C_2H_5OH(l) + O_2(g) \rightarrow CO_2(g) + H_2O(g)$$

Balance the skeletal equation.

$$C_2H_5OH(l) + 3\ O_2(g) \rightarrow 2\ CO_2(g) + 3\ H_2O(g)$$

SKILLBUILDER 7.15   Classifying Chemical Reactions According to What Atoms Do

Classify each of the following reactions as a synthesis, decomposition, single-displacement reaction, or double-displacement reaction.

a) $2\ Al(s) + 2\ H_3PO_4(aq) \rightarrow 2\ AlPO_4(aq) + 3\ H_2(g)$
b) $CuSO_4(aq) + 2\ KOH(aq) \rightarrow Cu(OH)_2(s) + K_2SO_4(aq)$
c) $2\ K(s) + Br_2(l) \rightarrow 2\ KBr(s)$
d) $CuCl_2(aq) \xrightarrow{\text{electrical current}} Cu(s) + Cl_2(g)$

Solution:
a) Single displacement; Al displaces H in H_3PO_4.
b) Double displacement; Cu and K switch places to form two new compounds.
c) Synthesis; a more complex substance forms from two simpler ones.
d) Decomposition; a complex substance decomposes into simpler ones.

SELF-TEST QUESTIONS

A. Match the following terms with the phrases below.

acid-base reaction	net ionic equation
balanced equation	oxidation-reduction (redox) reaction
chemical equation	precipitation reaction
chemical reaction	single-displacement reaction
complete ionic equation	solubility rules
double-displacement reaction	strong electrolyte solution
gas evolution reaction	synthesis reaction

1. Change of one or more substances into different substances
2. Reaction that occurs in a liquid and forms a gas
3. Reaction in which electrons are transferred from one substance to another
4. Reaction that forms one or more solid substances in water
5. Representation of a chemical reaction using chemical formulas

Chemical Reactions

6. Chemical equation with coefficients such that the number of each type of atom on each side of the equation is equal
7. Solution that contains a dissolved ionic compound
8. Empirical rules deduced from observations on the ability of many compounds to dissolve in water
9. Chemical equation showing all of the species as they are actually present in solution
10. Equation showing only the species that actually change during the reaction
11. Reaction between an acid and a base
12. Reaction in which simpler substances combine to form a more complex substance
13. Reaction in which one element displaces another in a compound
14. Reaction in which two elements or groups of elements in two different compounds exchange places to form two new compounds

B. **True/False**

1. Emission of light is evidence for a chemical reaction.
2. Chemical equations are balanced by placing coefficients in front of chemical formulas.
3. A compound is insoluble if it dissolves in water.
4. Whenever two substances are mixed, a chemical reaction occurs.
5. Spectator ions do not participate in chemical reactions.
6. Acid-base reactions generally form water and a salt.
7. Reduction is loss of electrons.
8. Combustion is a redox reaction.
9. Synthesis reactions involve combination of simpler substances to form more complex substances.
10. In a single displacement reaction, two elements or groups of elements in two different compounds exchange places to form two new compounds.

C. **Multiple Choice**

1. Which of the following provides evidence for a chemical reaction?
    a) A liquid freezes upon cooling
    b) A liquid evaporates upon heating
    c) A solid melts upon heating
    d) A substance is added to a solution and a gas forms

2. When the equation $C_4H_{10}(g) + O_2(g) \rightarrow CO_2(g) + H_2O(g)$ is balanced, the coefficient of O_2 is
    a) 4
    b) 8
    c) 9
    d) 13

Chapter 7

3. Which of the following compounds is insoluble in water?
   a) KCl
   b) Ca(NO$_3$)$_2$
   c) BaCO$_3$
   d) NH$_4$Br

4. What is the formula of the precipitate that forms when aqueous solutions of K$_3$PO$_4$ and BaBr$_2$ are mixed?
   a) KBr
   b) K$_3$Br$_2$
   c) BaPO$_4$
   d) Ba$_3$(PO$_4$)$_2$

5. In the net ionic equation for the reaction that occurs upon mixing aqueous solutions of NaCl and AgNO$_3$, the spectator ions are
   a) Na$^+$ and NO$_3^-$
   b) Na$^+$ and Cl$^-$
   c) Ag$^+$ and NO$_3^-$
   d) Ag$^+$ and Cl$^-$

6. When the equation NaOH(aq) + H$_3$PO$_4$(aq) → Na$_3$PO$_4$(s) + H$_2$O(l) is balanced, the coefficient of water is
   a) 2
   b) 3
   c) 4
   d) 6

7. Which of the following is a synthesis reaction?
   a) 2NaI(aq) + Cl$_2$(g) → 2NaCl(aq) + I$_2$(s)
   b) SF$_4$(g) + F$_2$(g) → SF$_6$(g)
   c) Na$_2$SO$_4$(aq) + CaI$_2$(aq) → CaSO$_4$(s) + 2NaI(aq)
   d) 2H$_2$O$_2$(l) → 2H$_2$O(l) + O$_2$(g)

8. Which of the following is a decomposition reaction?
   a) N$_2$O$_5$(s) + H$_2$O(l) → 2HNO$_3$(aq)
   b) H$_2$O$_2$(aq) + 2HI(aq) → I$_2$(s) + H$_2$O(l)
   c) MgCO$_3$(s) $\xrightarrow{\text{heat}}$ MgO(s) + CO$_2$(g)
   d) SO$_3$(g) + H$_2$O(l) → H$_2$SO$_4$(aq)

9. Which of the following is a redox reaction?
   a) 2Al(s) + 3Cl$_2$(g) → 2AlCl$_3$(s)
   b) 2HBr(aq) + LiCO$_3$ → LiBr(aq) + H$_2$O(l) + CO$_2$(g)
   c) Mg(OH)$_2$(s) → MgO(s) + H$_2$O(l)
   d) MnS(s) + 2HCl(aq) → MnCl$_2$(aq) + H$_2$S(g)

Chemical Reactions

10. Which of the following reactions is a single displacement reaction?

a) $H_2O(l) \xrightarrow{\text{electrical current}} H_2(g) + O_2(g)$

b) $Zn(s) + CuSO_4(aq) \rightarrow ZnSO_4(aq) + Cu(s)$

c) $HC_2H_3O_2(aq) + KOH(aq) \rightarrow KC_2H_3O_2(aq) + H_2O(l)$

d) $CaO(s) + CO_2(g) \rightarrow CaCO_3(s)$

D. Crossword Puzzle

ACROSS	DOWN
1. Type of reaction in which a complex substance decomposes to form simpler substances	2. Type of reaction in which a substance reacts with oxygen, emits heat, and forms one or more oxygen-containing compounds
3. Type of chemical equation showing the complete neutral formulas for every compound in a reaction	4. Light in the region of the electromagnetic spectrum that is higher in energy than visible light
6. Type of solution in which a substance is dissolved in water	5. Ionic compound
7. Type of ions that do not participate in a reaction	7. Type of compound that dissolves in water
9. Type of compound that does not dissolve in water	8. Solid formed upon mixing two aqueous solutions
10. Substances on the right side of a chemical equation	
11. Substances on the left side of a chemical equation	

ANSWERS TO SELF-TEST QUESTIONS

A. Matching
1. chemical reaction 2. gas evolution reaction 3. oxidation-reduction (redox) reaction 4. precipitation reaction 5. chemical equation 6. balanced equation 7. strong electrolyte solution 8. solubility rules 9. complete ionic equation 10. net ionic equation 11. acid-base reaction 12. synthesis reaction 13. single-displacement reaction 14. double-displacement reaction

B. True/False
1. T 2. T 3. F 4. F 5. T 6. T 7. F 8. T 9. T 10. F

C. Multiple Choice
1. d 2. d 3. c 4. d 5. a 6. b 7. b 8. c 9. a 10. b

Chemical Reactions

D. Crossword Puzzle

Across:
1. DECOMPOSITION
3. MOLECULAR
6. AQUEOUS
7. SPECTATOR
9. INSOLUBLE
10. PRODUCTS
11. REACTANTS

Down:
1. (not applicable)
2. COMBUSTION
3. MBS... / MOLES... (MBSTIO – from grid: M,O,B,S,T,I,O — appears as MOBSTIO)
4. ULTRAVIOLET
5. SALT
7. SOLUBLE
8. PRECIPITATE

Chapter 7

Quantities in Chemical Reactions   8

CHAPTER OVERVIEW

Chapter 8 concentrates on stoichiometry. Mole-to-mole and mass-to-mass conversions are presented. The concepts of limiting reactants, theoretical yield, and percent yield are explained.

CHAPTER OBJECTIVES

After reading and studying the text, students should be able to:

1. Use balanced chemical equations to perform mole-to-mole conversions.
2. Use balanced chemical equations to perform mass-to-mass conversions.
3. Find the limiting reactant, given a balanced chemical equation and the mass of each reactant.
4. Calculate the theoretical yield of a reaction, given a balanced chemical equation and the mass of each reactant.
5. Calculate the percent yield from the actual yield and theoretical yield of a reaction.

CHAPTER IN REVIEW

- Greenhouse gases are gases in the earth's atmosphere that affect the balance between incoming sunlight, which warms the earth, and outgoing heat loss to space. Carbon dioxide is the earth's most significant greenhouse gas in terms of its contribution to climate.

- Global warming is an increase in the earth's average temperature. Scientists are concerned that the measured increase in carbon dioxide, primarily caused by fossil fuel combustion, is causing the observed increase in the earth's average temperature.

- Balanced chemical equations give quantitative relationships between the amounts of reactants and products.

- To convert between the number of moles of one substance and the number of moles of another substance, use the mole-to-mole conversion factor that comes from the balanced equation.

- To convert from the mass of one substance to the mass of another substance in a chemical equation, convert the given mass to number of moles using molar mass, convert from number of moles of that substance to the number of moles of the other substance using

Chapter 8

the mole-to-mole conversion factor from the balanced equation, then convert from the number of moles to the mass of the substance using its molar mass.

- The limiting reactant is the reactant that makes the least amount of product. The limiting reactant is completely consumed in the chemical reaction.

- The theoretical yield is the amount of product that can be made in a chemical reaction based on the amount of the limiting reactant.

- The actual yield is the amount of product actually produced by a chemical reaction.

- Percent yield is the percentage of the theoretical yield that was actually obtained.

- To find the limiting reactant, calculate the amount of product that can be made from each reactant. The reactant that makes the least amount of product is the limiting reactant.

SKILLBUILDER PROBLEMS AND SOLUTIONS

SKILLBUILDER 8.1   **Mole-to-Mole Conversions**

Water is formed when hydrogen gas reacts explosively with oxygen gas according to the following balanced equation.

$$O_2(g) + 2\ H_2(g) \rightarrow 2\ H_2O(g)$$

How many moles of H_2O result from the complete reaction of 24.6 mol of O_2? Assume that there is more than enough H_2.

Given: 24.6 mol O_2

Find: mol H_2O

Conversion Factor: The conversion factor comes from the balanced chemical equation.

$$1\ \text{mol}\ O_2 \equiv 2\ \text{mol}\ H_2O$$

Solution Map:

mol O_2 → mol H_2O

$$\frac{2\ \text{mol}\ H_2O}{1\ \text{mol}\ O_2}$$

Solution:

$$24.6 \text{ mol } O_2 \times \frac{2 \text{ mol } H_2O}{1 \text{ mol } O_2} = 49.2 \text{ mol } H_2O$$

SKILLBUILDER 8.2   Mass-to-Mass Conversions

Magnesium hydroxide, the active ingredient in milk of magnesia, neutralizes stomach acid, primarily HCl, according to the following reaction.

$$Mg(OH)_2(aq) + 2 \text{ HCl}(aq) \rightarrow 2 \text{ H}_2O(l) + MgCl_2(aq)$$

How much HCl in grams can be neutralized by 5.50 g Mg(OH)$_2$?

Given: 5.50 g Mg(OH)$_2$

Find: g HCl

Conversion Factors:

1 mol Mg(OH)$_2$ ≡ 2 mol HCl (from balanced chemical equation)
Molar mass Mg(OH)$_2$ = 58.33 g/mol
Molar mass HCl = 36.46 g/mol

Solution Map:

g Mg(OH)$_2$ → mol Mg(OH)$_2$ → mol HCl → g HCl

$$\frac{1 \text{ mol Mg(OH)}_2}{58.33 \text{ g Mg(OH)}_2} \quad \frac{2 \text{ mol HCl}}{1 \text{ mol Mg(OH)}_2} \quad \frac{36.46 \text{ g HCl}}{1 \text{ mol HCl}}$$

Solution:

$$5.50 \text{ g Mg(OH)}_2 \times \frac{1 \text{ mol Mg(OH)}_2}{58.33 \text{ g Mg(OH)}_2} \times \frac{2 \text{ mol HCl}}{1 \text{ mol Mg(OH)}_2} \times \frac{36.46 \text{ g HCl}}{1 \text{ mol HCl}}$$

= 6.88 g HCl

SKILLBUILDER 8.3   Mass-to-Mass Conversions

A component of acid rain is sulfuric acid, which forms when SO$_2$, also a pollutant, reacts with oxygen and rainwater according to the following reaction.

$$2 \text{ SO}_2(g) + O_2(g) + 2 \text{ H}_2O(l) \rightarrow 2 \text{ H}_2SO_4(aq)$$

Assuming that there is plenty of O$_2$ and H$_2$O, how much H$_2$SO$_4$ in kilograms forms from

Chapter 8

2.6×10^3 kg of SO_2?

Given: 2.6×10^3 kg of SO_2

Find: kg H_2SO_4

Conversion Factors:

2 mol SO_2 ≡ 2 mol H_2SO_4 (from balanced chemical equation)
Molar mass SO_2 = 64.07 g/mol
Molar mass H_2SO_4 = 98.09 g/mol
1 kg = 1000 g

Solution Map:

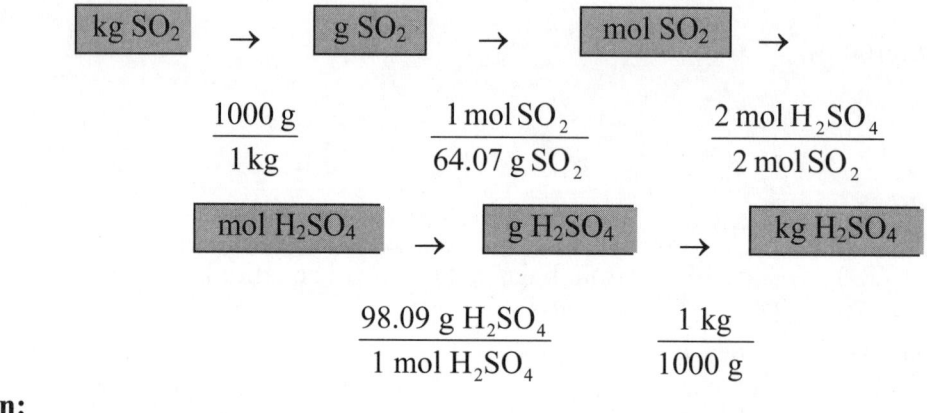

kg SO_2 → g SO_2 → mol SO_2 →

$$\frac{1000 \text{ g}}{1 \text{ kg}} \quad \frac{1 \text{ mol } SO_2}{64.07 \text{ g } SO_2} \quad \frac{2 \text{ mol } H_2SO_4}{2 \text{ mol } SO_2}$$

mol H_2SO_4 → g H_2SO_4 → kg H_2SO_4

$$\frac{98.09 \text{ g } H_2SO_4}{1 \text{ mol } H_2SO_4} \quad \frac{1 \text{ kg}}{1000 \text{ g}}$$

Solution:

$$2.6 \times 10^3 \text{ kg } SO_2 \times \frac{1000 \text{ g}}{1 \text{ kg}} \times \frac{1 \text{ mol } SO_2}{64.07 \text{ g } SO_2} \times \frac{2 \text{ mol } H_2SO_4}{2 \text{ mol } SO_2}$$

$$\times \frac{98.09 \text{ g } H_2SO_4}{1 \text{ mol } H_2SO_4} \times \frac{1 \text{ kg}}{1000 \text{ g}} = 4.0 \times 10^3 \text{ kg } H_2SO_4$$

SKILLBUILDER 8.4   **Limiting Reactant and Theoretical Yield from Initial Moles of Reactants**

Consider the following reaction.

$$2 \text{ Na}(s) + F_2(g) \rightarrow 2 \text{ NaF}(s)$$

If you begin with 4.8 mol of sodium and 2.6 mol of fluorine, what is the limiting reactant and theoretical yield of NaF in moles?

Given: 4.8 mol Na, 2.6 mol F_2

Find: limiting reactant, theoretical yield

Quantities in Chemical Reactions

Conversion Factors:

From the balanced equation we have

    2 mol Na ≡ 2 mol NaF
    1 mol F_2 ≡ 2 mol NaF

The solution map shows how to get from moles of each reactant to mol NaF. The reactant that makes the *least amount of NaF* is the limiting reactant.

Solution Map:

    mol Na → mol NaF

$$\frac{2 \text{ mol NaF}}{2 \text{ mol Na}}$$

    mol F_2 → mol NaF

$$\frac{2 \text{ mol NaF}}{1 \text{ mol } F_2}$$

Solution:

$$4.8 \text{ mol Na} \times \frac{2 \text{ mol NaF}}{2 \text{ mol Na}} = 4.8 \text{ mol NaF}$$

$$2.6 \text{ mol } F_2 \times \frac{2 \text{ mol NaF}}{1 \text{ mol } F_2} = 5.2 \text{ mol NaF}$$

Since the 4.8 mol Na makes the least amount of NaF, Na is the limiting reactant. The theoretical yield is 4.8 mol of NaF.

SKILLBUILDER 8.5   **Finding Limiting Reactant and Theoretical Yield**

Ammonia can by synthesized by the following reaction.

$$3 H_2(g) + N_2(g) \rightarrow 2 NH_3(g)$$

What is the maximum amount of ammonia that can be synthesized from 25.2 g of N_2 and 8.42 g of H_2?

Given: 25.2 g N_2, 8.42 g H_2

Chapter 8

Find: maximum amount of NH₃ (theoretical yield)

Conversion Factors:

3 mol H₂ ≡ 2 mol NH₃
1 mol N₂ ≡ 2 mol NH₃
Molar mass H₂ = 2.02 g/mol
Molar mass N₂ = 28.02 g/mol
Molar mass NH₃ = 17.04 g/mol

We find the limiting reactant by calculating how much product can be made from each reactant. Since we are given the initial amounts in grams, we must first convert to moles. After we convert to moles of product, we convert to grams of product. The reactant that makes the *least amount of product* is the limiting reactant.

Solution Map:

g H₂ → mol H₂ → mol NH₃ → g NH₃

$$\frac{1 \text{ mol } H_2}{2.02 \text{ g } H_2} \quad \frac{2 \text{ mol } NH_3}{3 \text{ mol } H_2} \quad \frac{17.04 \text{ g } NH_3}{1 \text{ mol } NH_3}$$

g N₂ → mol N₂ → mol NH₃ → g NH₃

$$\frac{1 \text{ mol } N_2}{28.02 \text{ g } N_2} \quad \frac{2 \text{ mol } NH_3}{1 \text{ mol } N_2} \quad \frac{17.04 \text{ g } NH_3}{1 \text{ mol } NH_3}$$

Solution:

$$8.42 \text{ g } H_2 \times \frac{1 \text{ mol } H_2}{2.02 \text{ g } H_2} \times \frac{2 \text{ mol } NH_3}{3 \text{ mol } H_2} \times \frac{17.04 \text{ g } NH_3}{1 \text{ mol } NH_3} = 47.4 \text{ g } NH_3$$

$$25.2 \text{ g } N_2 \times \frac{1 \text{ mol } N_2}{28.02 \text{ g } N_2} \times \frac{2 \text{ mol } NH_3}{1 \text{ mol } N_2} \times \frac{17.04 \text{ g } NH_3}{1 \text{ mol } NH_3} = 30.7 \text{ g } NH_3$$

We have enough H₂ to make 47.4 g NH₃ and enough N₂ to make 30.7 g NH₃. Therefore, N₂ is the limiting reactant and the maximum amount of ammonia that we can possibly make is 30.7 g, the theoretical yield.

SKILLBUILDER PLUS

Ammonia can by synthesized by the following reaction.

Quantities in Chemical Reactions

$$3\ H_2(g)\ +\ N_2(g)\ \rightarrow\ 2\ NH_3(g)$$

What is the maximum amount of ammonia in kilograms that can be synthesized from 5.22 kg of H_2 and 31.5 kg of N_2?

Given: 5.22 kg H_2, 31.5 kg N_2

Find: maximum amount of NH_3 (theoretical yield)

Conversion Factors:

1000 g = 1 kg
3 mol H_2 ≡ 2 mol NH_3
1 mol N_2 ≡ 2 mol NH_3
Molar mass H_2 = 2.20 g/mol
Molar mass N_2 = 28.02 g/mol
Molar mass NH_3 = 17.04 g/mol

Solution Map:

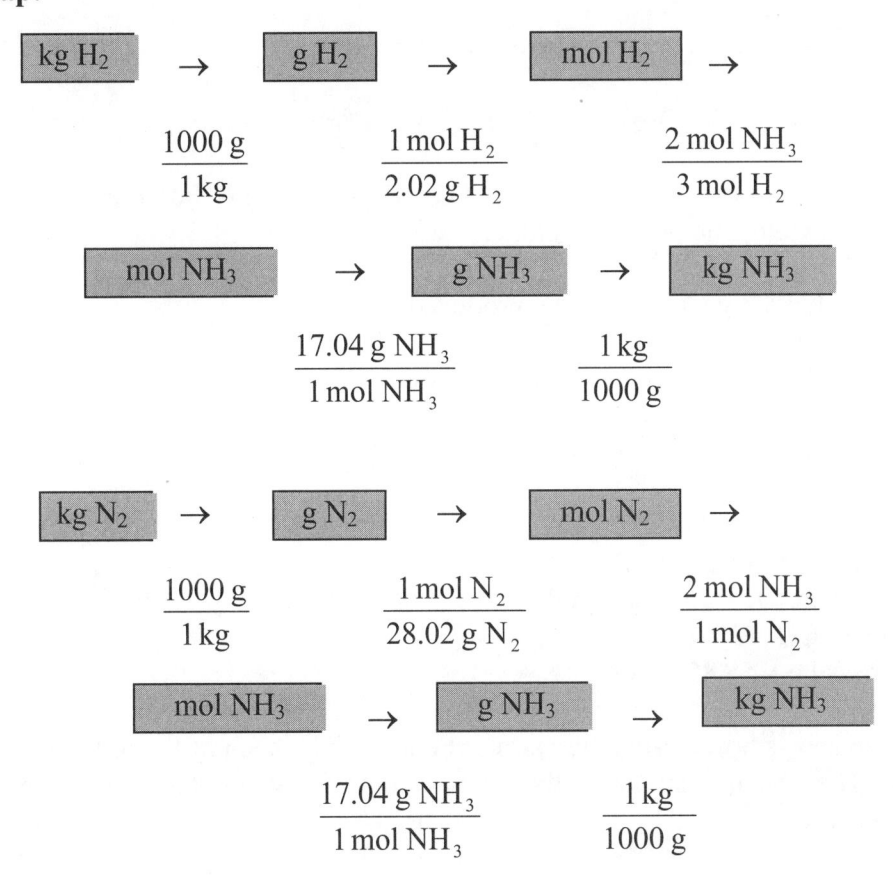

$$\boxed{kg\ H_2} \rightarrow \boxed{g\ H_2} \rightarrow \boxed{mol\ H_2} \rightarrow$$

$$\frac{1000\ g}{1\ kg} \qquad \frac{1\ mol\ H_2}{2.02\ g\ H_2} \qquad \frac{2\ mol\ NH_3}{3\ mol\ H_2}$$

$$\boxed{mol\ NH_3} \rightarrow \boxed{g\ NH_3} \rightarrow \boxed{kg\ NH_3}$$

$$\frac{17.04\ g\ NH_3}{1\ mol\ NH_3} \qquad \frac{1\ kg}{1000\ g}$$

$$\boxed{kg\ N_2} \rightarrow \boxed{g\ N_2} \rightarrow \boxed{mol\ N_2} \rightarrow$$

$$\frac{1000\ g}{1\ kg} \qquad \frac{1\ mol\ N_2}{28.02\ g\ N_2} \qquad \frac{2\ mol\ NH_3}{1\ mol\ N_2}$$

$$\boxed{mol\ NH_3} \rightarrow \boxed{g\ NH_3} \rightarrow \boxed{kg\ NH_3}$$

$$\frac{17.04\ g\ NH_3}{1\ mol\ NH_3} \qquad \frac{1\ kg}{1000\ g}$$

Chapter 8

Solution:

$$5.22 \text{ kg H}_2 \times \frac{1000 \text{ g}}{1 \text{ kg}} \times \frac{1 \text{ mol H}_2}{2.02 \text{ g H}_2} \times \frac{2 \text{ mol NH}_3}{3 \text{ mol H}_2} \times \frac{17.04 \text{ g NH}_3}{1 \text{ mol NH}_3} \times \frac{1 \text{ kg}}{1000 \text{ g}}$$

$$= 29.4 \text{ kg NH}_3$$

$$31.5 \text{ kg N}_2 \times \frac{1000 \text{ g}}{1 \text{ kg}} \times \frac{1 \text{ mol N}_2}{28.02 \text{ g N}_2} \times \frac{2 \text{ mol NH}_3}{1 \text{ mol N}_2} \times \frac{17.04 \text{ g NH}_3}{1 \text{ mol NH}_3} \times \frac{1 \text{ kg}}{1000 \text{ g}}$$

$$= 38.3 \text{ kg NH}_3$$

We have enough H_2 to make 29.4 kg NH_3 and enough N_2 to make 38.3 kg NH_3. Therefore, H_2 is the limiting reactant and the maximum amount of ammonia that we can possibly make is 29.4 kg, the theoretical yield.

SKILLBUILDER 8.6 **Finding Limiting Reactant, Theoretical Yield, and Percent Yield**

The following reaction is used to obtain iron from iron ore.

$$Fe_2O_3(s) + 3 \, CO(g) \rightarrow 2 \, Fe(s) + 3 \, CO_2(g)$$

The reaction of 185 g of Fe_2O_3 with 95.3 g of CO produces 87.4 g of Fe. Find the limiting reactant, theoretical yield, and percent yield.

Given: 185 g Fe_2O_3, 95.3 g CO, 87.4 g Fe produced

Find: limiting reactant, theoretical yield, percent yield

Conversion Factors:

1 mol Fe_2O_3 ≡ 2 mol Fe
3 mol CO ≡ 2 mol Fe
Molar mass Fe_2O_3 = 159.70 g/mol
Molar mass CO = 28.01 g/mol
Molar mass Fe = 55.85 g/mol

The solution map shows how to find the mass of Fe formed by each of the initial masses of Fe_2O_3 and CO. The reactant that makes the least amount of product is the limiting reactant and determines the theoretical yield.

Quantities in Chemical Reactions

Solution Map:

g Fe₂O₃ → mol Fe₂O₃ → mol Fe → g Fe

$$\frac{1 \text{ mol Fe}_2\text{O}_3}{159.70 \text{ g Fe}_2\text{O}_3} \qquad \frac{2 \text{ mol Fe}}{1 \text{ mol Fe}_2\text{O}_3} \qquad \frac{55.85 \text{ g Fe}}{1 \text{ mol Fe}}$$

g CO → mol CO → mol Fe → g Fe

$$\frac{1 \text{ mol CO}}{28.01 \text{ g CO}} \qquad \frac{2 \text{ mol Fe}}{3 \text{ mol CO}} \qquad \frac{55.85 \text{ g Fe}}{1 \text{ mol Fe}}$$

Solution:

$$185 \text{ g Fe}_2\text{O}_3 \times \frac{1 \text{ mol Fe}_2\text{O}_3}{159.70 \text{ g Fe}_2\text{O}_3} \times \frac{2 \text{ mol Fe}}{1 \text{ mol Fe}_2\text{O}_3} \times \frac{55.85 \text{ g Fe}}{1 \text{ mol Fe}} = 129 \text{ g Fe}$$

$$95.3 \text{ g CO} \times \frac{1 \text{ mol CO}}{28.01 \text{ g CO}} \times \frac{2 \text{ mol Fe}}{3 \text{ mol CO}} \times \frac{55.85 \text{ g Fe}}{1 \text{ mol Fe}} = 127 \text{ g Fe}$$

Since CO makes the least amount of product, CO is the limiting reactant. The theoretical yield is simply the amount of product made by the limiting reactant.

Theoretical yield = 127 g Fe

The percent yield is

$$\text{Percent yield} = \frac{\text{Actual yield}}{\text{Theoretical yield}} \times 100\%$$

$$= \frac{87.4 \text{ g}}{127 \text{ g}} \times 100\% = 68.8\%$$

SELF-TEST QUESTIONS

A. Match the following terms with the phrases below.

actual yield
global warming
greenhouse gases

limiting reactant
percent yield
theoretical yield

Chapter 8

1. Gases in the earth's atmosphere that affect the balance between incoming sunlight, which warms the earth, and outgoing heat lost to space
2. Increase in the earth's average temperature
3. Reactant that limits the amount of product in a chemical reaction
4. Amount of product that can be made in a chemical reaction based on the amount of limiting reactant
5. Amount of product actually produced by a chemical reaction
6. Percentage of the theoretical yield actually obtained

B. True/False

1. The main cause of rising carbon dioxide levels in the atmosphere is fossil fuel combustion.
2. The numerical relationships between chemical quantities in a balanced chemical equation is called reaction stoichiometry.
3. The symbol $\equiv$ means "is equivalent to."
4. Coefficients in balanced equations give mass-to-mass relationships.
5. The sum of the coefficients of reactants must equal the sum of the coefficients of products in a balanced chemical equation.
6. The limiting reactant is the reactant present in the smallest amount of moles.
7. The limiting reactant is completely consumed in a chemical reaction.
8. The theoretical yield is based on the amount of limiting reactant.
9. The actual yield of a reaction can be calculated from the quantities of reactants used.
10. Percent yield is calculated by dividing the theoretical yield by the actual yield, then multiplying by 100.

C. Multiple Choice

1. For the reaction shown, how many moles of HCl react with 0.278 mol of Al?

$$2\ Al(s) + 6\ HCl\ (aq) \rightarrow 2\ AlCl_3(aq) + 3\ H_2(g)$$

   a) 0.278 mol
   b) 0.834 mol
   c) 0.927 mol
   d) 1.67 mol

2. For the reaction shown, how many moles of CO_2 are formed by the complete reaction of 0.0534 mol of C_2H_6O?

$$C_2H_6O(g) + 3\ O_2(g) \rightarrow 2\ CO_2(g) + 3\ H_2O(g)$$

   a) 0.0535 mol
   b) 0.107 mol
   c) 0.160 mol
   d) 0.320 mol

3. Consider the following reaction.

$$4\ NH_3(g) + 3\ O_2(g) \rightarrow 2\ N_2(g) + 6\ H_2O(g)$$

How many grams of O_2 are needed to react with 9.92 g of NH_3?

a) 4.66 g
b) 14.0 g
c) 18.6 g
d) 55.9 g

4. For the reaction shown, how many grams of Al form if 6.16 g of Al_2O_3 completely react with H_2?

$$Al_2O_3(s) + 3\ H_2(g) \rightarrow 2\ Al(s) + 3\ H_2O(g)$$

a) 0.121 g
b) 1.63 g
c) 3.08 g
d) 3.26 g

5. Consider the following equation for the reaction of zinc with hydrochloric acid.

$$Zn(s) + 2\ HCl(aq) \rightarrow ZnCl_2(aq) + H_2(g)$$

If 0.345 mol of Zn is combined with 0.580 mol of HCl, what is the theoretical number of moles of $ZnCl_2$ that can form?

a) 0.290 mol
b) 0.345 mol
c) 0.580 mol
d) 0.925 mol

6. For the reaction shown below, 0.145 mol HCl is combined with 0.112 mol $Ba(OH)_2$. Identify the limiting reactant.

$$2\ HCl(aq) + Ba(OH)_2(aq) \rightarrow BaCl_2(aq) + 2\ H_2O(l)$$

a) HCl
b) $Ba(OH)_2$
c) $BaCl_2$
d) H_2O

Chapter 8

7. Consider the following reaction.

$$Cu(s) + Pb(NO_3)_2(aq) \rightarrow Cu(NO_3)_2(aq) + Pb(s)$$

If 2.43 g of Cu is reacted with 8.13 g of Pb(NO$_3$)$_2$, what is the theoretical yield of Cu(NO$_3$)$_2$?

a) 4.60 g
b) 7.17 g
c) 10.6 g
d) 11.8 g

8. For the reaction shown below, 8.02 g of KI are combined with 9.84 g of Pb(NO$_3$)$_2$. Identify the limiting reactant.

$$2\ KI(aq) + Pb(NO_3)_2(aq) \rightarrow 2\ KNO_3(aq) + PbI_2(s)$$

a) KI
b) Pb(NO$_3$)$_2$
c) KNO$_3$
d) PbI$_2$

9. Consider the reaction below. A solution containing 4.40 g of CaCl$_2$ is combined with a solution containing 5.60 g of Na$_3$PO$_4$. Identify the reactant present in excess.

$$3\ CaCl_2(aq) + 2\ Na_3PO_4(aq) \rightarrow Ca_3(PO_4)_2(s) + 6\ NaCl(aq)$$

a) CaCl$_2$
b) Na$_3$PO$_4$
c) Ca$_3$(PO$_4$)$_2$
d) NaCl

10. Consider the following reaction.

$$2\ Li(s) + 2\ H_2O(l) \rightarrow 2\ LiOH(aq) + H_2(g)$$

If 0.568 g of Li reacts with excess H$_2$O and 0.0616 g of H$_2$ is collected, what is the percent yield?

a) 18.6%
b) 37.3%
c) 67.0%
d) 74.5%

Quantities in Chemical Reactions

D. Crossword Puzzle

ACROSS

3. Type of chemical equation that provides mole-to-mole conversion factors
4. A component of gasoline
5. Type of burner commonly used in chemistry laboratories
6. Numerical relationship between chemical quantities in a balanced chemical equation

DOWN

1. Burning of fuel
2. Gasoline additive made from fermentation of grain

ANSWERS TO SELF-TEST QUESTIONS

A. Matching
1. greenhouse gases 2. global warming 3. limiting reactant 4. theoretical yield 5. actual yield
6. percent yield

B. True/False
1. T 2. T 3. T 4. F 5. F 6. F 7. T 8. T 9. F 10. F

C. Multiple Choice
1. b 2. b 3. b 4. d 5. a 6. a 7. a 8. a 9. b 10. d

Chapter 8

D. Crossword Puzzle

	1	2	3	4	5	6	7	8	9	10	11	12	13
1	¹C						²E		³B	A	L	A	N	C	E	D
2	⁴O	C	T	A	N	E		T						
3	M						H						
4	⁵B	U	N	S	E	N		A						
5	U						N						
6	⁶S	T	O	I	C	H	I	O	M	E	T	R	Y	
7	T						O						
8	I						L						
9	O												
10	N												

118

Electrons in Atoms and the Periodic Table

CHAPTER OVERVIEW

Chapter 9 explores electromagnetic radiation in conjunction with the electronic structure of the atom. Electron configurations and orbital diagrams for atoms and ions are related to the periodic table. Periodic trends in ionization energy, atomic size, and metallic character are discussed.

CHAPTER OBJECTIVES

After reading and studying the text, students should be able to:

1. State the relationship between the wavelength, frequency, and energy of light.
2. List the regions of the electromagnetic spectrum and arrange them in order of increasing wavelength, frequency, or energy per photon.
3. Summarize the Bohr model of the atom.
4. Explain the difference between a Bohr orbit and a quantum mechanical orbital.
5. Describe the difference between the ground state and an excited state of an atom.
6. List the four possible subshells in the quantum mechanical model.
7. State the maximum number of electrons that can be contained in each subshell of the quantum mechanical model.
8. List all of the quantum mechanical orbitals in order of increasing energy through $6s$.
9. State the Pauli exclusion principle.
10. State Hund's rule.
11. Write electron configurations for atoms and monoatomic ions.
12. Draw orbital diagrams for atoms and monoatomic ions.
13. Differentiate between core electrons and valence electrons.
14. Write the electron configuration of an element based on its position in the periodic table.
15. Use the periodic table to arrange elements in order of increasing atomic size.
16. Use the periodic table to arrange elements in order of increasing ionization energy.
17. Use the periodic table to arrange elements in order of increasing metallic character.

CHAPTER IN REVIEW

- Light is electromagnetic radiation, a form of energy that exhibits both wave-like and particle-like behaviors. It travels through space at a speed of 3.0×10^8 m/s.

Chapter 9

- Particles of light are called photons. The wave nature of light is characterized by its wavelength (λ), the distance between adjacent wave crests. The frequency (ν) of light is the number of cycles that pass a stationary point in one second.

- The shorter the wavelength of light, the higher its energy and frequency.

- The electromagnetic spectrum includes all wavelengths of electromagnetic radiation. In order of increasing energy, the regions of the electromagnetic spectrum are radiowaves, microwaves, infrared radiation, visible light, ultraviolet radiation, X-rays, and gamma rays.

- The Bohr model of the atom explains the emission spectrum of hydrogen. According to the Bohr model, electrons occupy circular orbits at specific fixed distances from the nucleus. Each orbit is specified by a quantum number, n, which also specifies the orbit's energy. As the value of n increases, the distance of the orbit from the nucleus increases, and the energy of the orbit increases. When an electron transitions to a higher energy orbit, it absorbs a quantum of energy; when it transitions to a lower energy orbit it emits a quantum of energy. The energy (and therefore the wavelength) of the photons absorbed or emitted corresponds to the energy difference between the orbits.

- The quantum mechanical model of the atom predicts and explains chemical properties of all atoms. According to the quantum mechanical model, orbitals show the probability of where the electron will be found when the atom is probed. Orbitals are specified by the principal quantum number (n) and a letter. The principal quantum number is an integer (1, 2, 3, 4 …) which specifies the principal shell of the orbital. The letter (s, p, d, or f) specifies the subshell of the orbital. The number of subshells in a given principal shell is equal to the value of n.

- Orbitals increase in energy in the order 1s 2s 2p 3s 3p 4s 3d 4p 5s 4d 5p 6s.

- Electrons occupy orbitals in order of increasing energy.

- The Pauli exclusion principle states that each orbital can hold a maximum of two electrons.

- Hund's rule states that electrons occupy orbitals of identical energy singly before pairing.

- Electron configurations and orbital diagrams show the occupation of orbitals by electrons for a single atom.

- An atom's electron configuration can be deduced from the element's position on the periodic table.

- Valence electrons are the electrons in the outermost principal shell. Core electrons are electrons that are not in the outermost principal shell.

Electrons in Atoms and the Periodic Table

- The chemical properties of elements are largely determined by the number of valence electrons they contain.

- Atomic size is the distance from an atom's nucleus to its outermost electrons. Atomic size decreases going from left to right across the periodic table and increases going down a column of the periodic table.

- Ionization energy is the energy required to remove an electron from an atom. Ionization energy increases from left to right across a row of the periodic table and decreases going down a column of the periodic table.

- Metallic character decreases going from left to right across the periodic table and increases going down a column of the periodic table.

SKILLBUILDER PROBLEMS AND SOLUTIONS

SKILLBUILDER 9.1   Wavelength, Energy, and Frequency

Arrange the following colors of visible light—green, red, and blue—in order of increasing:

a) wavelength
b) frequency
c) energy per photon

Solution:
a) We can examine Figure 9.4 in the text to see that blue light has the shortest wavelength, followed by green light and then red light:

   blue light, green light, red light

b) Since frequency and wavelength are inversely proportional—the longer the wavelength the shorter the frequency—ordering with respect to frequency is exactly the reverse of the ordering with respect to wavelength:

   red light, green light, blue light

c) Energy per photon decreases with increasing wavelength but increases with increasing frequency; therefore, the ordering with respect to energy per photon is the same as frequency:

   red light, green light, blue light

Chapter 9

SKILLBUILDER 9.2   **Electron Configurations**

Write electron configurations for each of the following elements.

a) Al
b) Br
c) Sr

Solution:

a) Aluminum has 13 electrons. We distribute two of these into the 1s orbital, two into the 2s orbital, six into the 2p orbitals, two into the 3s orbital, and one into a 3p orbital. The electron configuration is

   Al $1s^2 2s^2 2p^6 3s^2 3p^1$

We can write this more compactly as

   Al   [Ne] $3s^2 3p^1$

b) Bromine has 35 electrons. We distribute two of these into the 1s orbital, two into the 2s orbital, six into the 2p orbitals, two into the 3s orbital, six into the 3p orbitals, two into the 4s orbital, ten into the 3d orbitals, and five into the 4p orbitals. The electron configuration is

   Br $1s^2 2s^2 2p^6 3s^2 3p^6 4s^2 3d^{10} 4p^5$

We can write this more compactly as

   Br   [Ar] $4s^2 3d^{10} 4p^5$

c) Strontium has 38 electrons. We distribute two of these into the 1s orbital, two into the 2s orbital, six into the 2p orbitals, two into the 3s orbital, six into the 3p orbitals, two into the 4s orbital, ten into the 3d orbitals, six into the 4p orbitals, and two into the 5s orbital. The electron configuration is

   Sr $1s^2 2s^2 2p^6 3s^2 3p^6 4s^2 3d^{10} 4p^6 5s^2$

We can write this more compactly as

   Sr   [Kr] $5s^2$

SKILLBUILDER PLUS

Write electron configurations for each of the following ions.

a) Al^{3+}
b) Cl^-
c) O^{2-}

Solution:
a) A neutral Al atom has 13 electrons. Al^{3+} has three fewer electrons, $13 - 3 = 10$ electrons. We distribute two of these into the 1s orbital, two into the 2s orbital, and six into the 2p orbitals. The electron configuration is the same as that for Ne.

$$Al^{3+} \quad 1s^2 2s^2 2p^6$$

b) A neutral chlorine atom has 17 electrons. Cl^- has one more electron, $17 + 1 = 18$ electrons. We distribute two of these into the 1s orbital, two into the 2s orbital, six into the 2p orbitals, two into the 3s orbital, and six into the 3p orbitals. The electron configuration is that same as that for Ar.

$$Cl^- \quad 1s^2 2s^2 2p^6 3s^2 3p^6$$

c) A neutral oxygen atom has 8 electrons. O^{2-} has two more electrons, $8 + 2 = 10$ electrons. We distribute two of these into the 1s orbital, two into the 2s orbital, and six into the 2p orbitals. The electron configuration is the same as that for Ne.

$$O^{2-} \quad 1s^2 2s^2 2p^6$$

SKILLBUILDER 9.3 **Writing Orbital Diagrams**

Write an orbital diagram for argon.

Solution:
Since argon has an atomic number of 18, it has 18 electrons. Draw a box for each orbital, putting the lowest energy orbital (1s) on the far left and proceeding to orbitals of higher energy to the right.

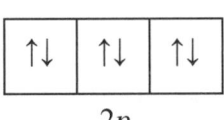

Ar [↑↓]   [↑↓]   [↑↓][↑↓][↑↓]   [↑↓]   [↑↓][↑↓][↑↓]
     1s   2s   2p   3s   3p

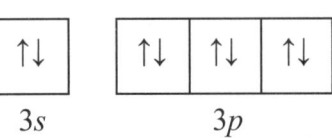

SKILLBUILDER 9.4 **Valence Electrons and Core Electrons**

Write an electron configuration for chlorine and identify the valence electrons and core electrons.

Chapter 9

Solution:
We write the electron configuration for chlorine by determining the total number of electrons from chlorine's atomic number (17) and then distribute them into the appropriate orbitals.

$$\text{Cl} \quad 1s^2 2s^2 2p^6 3s^2 3p^5$$

The valence electrons are those in the outermost principal shell. For chlorine, the outermost principal shell is the $n = 3$ shell, which contains 7 electrons (2 in the $3s$ orbital and 5 in the $3p$ orbitals). All other electrons are core electrons.

$$\text{Cl} \quad \underbrace{1s^2 2s^2 2p^6}_{\text{core electrons}} \quad \underbrace{3s^2 3p^5}_{\text{valence electrons}}$$

SKILLBUILDER 9.5   **Writing Electron Configurations from the Periodic Table**

Use the periodic table to determine the electron configuration for tin.

Solution:
The noble gas that precedes tin in the periodic table is krypton, so the inner electron configuration is [Kr]. Obtain the outer electron configuration by tracing the elements between Kr and Sn and assigning electrons to the appropriate orbitals:

$$\text{Sn} \quad [\text{Kr}]5s^2 4d^{10} 5p^2$$

SKILLBUILDER 9.6   **Atomic Size**

Choose the larger atom from each of the following pairs.

a) Pb or Po
b) Rb or Na
c) Sn or Bi
d) F or Se

Solution:
a) Pb atoms are larger than Po atoms because as we trace the path from Pb to Po on the periodic table, we move to the right within the same period. Atomic size decreases moving from left to right across a period.

b) Rb atoms are larger than Na atoms because as we trace the path from Rb to Na on the periodic table, we move up a column. Atomic size decreases going up a column.

c) Based on periodic properties alone, we cannot tell which atom is larger because as we trace the path from Sn to Bi on the periodic table, we move down a column (atomic size increases)

124

Electrons in Atoms and the Periodic Table

and then to the right across a period (atomic size decreases). These effects tend to cancel one another.

d) Se atoms are larger than F atoms because as we trace the path from F to Se on the periodic table, we move down a column (atomic size increases) and then to the left across a period (atomic size increases). These effects add together for an overall increase.

SKILLBUILDER 9.7 Ionization Energy

Choose the element with the higher ionization energy from each of the following pairs.

a) Mg or Sr
b) In or Te
c) C or P
d) F or S

Solution:
a) Mg has a higher ionization energy than Sr because as we trace the path from Mg to Sr on the periodic table, we move down a column. Ionization energy decreases going down a column.

b) Te has a higher ionization energy than In because as we trace the path from In to Te on the periodic table, we move to the right within the same period. Ionization energy increases going from left to right across a period.

c) Based on periodic properties alone, we cannot tell which has a higher ionization energy because as we trace the path from C to P, we go down a column (ionization energy decreases) and then to the right across a period (ionization energy increases). These effects tend to cancel

d) F has a higher ionization energy than S because as we trace the path from F to S on the periodic table, we move down a column (ionization energy decreases) and then to the left across the period (ionization energy decreases). These effects add together for an overall decrease.

SKILLBUILDER 9.8 Metallic Character

Choose the more metallic element from each of the following pairs.

a) Ge or In
b) Ga or Sn
c) P or Bi
d) B or N

Chapter 9

Solution:
a) In is more metallic than Ge because as we trace the path from Ge to In on the periodic table, we move down a column (metallic character increases) and then to the left across a period (metallic character increases). These effects add together for an overall increase.

b) Based on periodic properties alone, we cannot tell which is more metallic because as we trace the path from Ga to Sn, we go down a column (metallic character increases) and then to the right across a period (metallic character decreases). These effects tend to cancel.

c) Bi is more metallic than P because as we trace the path from P to Bi on the periodic table, we move down a column. Metallic character increases going down a column.

e) B is more metallic than N because as we trace the path from B to N on the periodic table, we move to the right within the same period. Metallic character decreases going from left to right across a period.

SELF-TEST QUESTIONS

A. Match the following terms with the phrases below.

atomic size	Pauli exclusion principle
Bohr model	principal quantum number
electromagnetic spectrum	quantum numbers
electron configuration	quantum-mechanical model
ground state	radiowaves
Hund's rule	ultraviolet light
orbital diagram	X-rays

1. Model of the atom that says that electrons exist in orbits at specific, fixed energies and specific, fixed distances from the nucleus
2. Model of the atom that invokes probability maps of where the electron is likely to be found
3. Energy spectrum ranging from radiowaves to gamma rays
4. Form of electromagnetic radiation with wavelengths between those of gamma rays and ultraviolet rays
5. Form of electromagnetic radiation with wavelengths between those of X-rays and visible light
6. Form of electromagnetic radiation with the longest wavelengths
7. Stable energy state of an atom
8. Numbers that specify the energy of each Bohr orbit
9. Number that represents the principal shell of an orbital
10. Representation of the occupation of orbitals by electrons for a particular atom that uses numbers, letters, and superscripts
11. Representation of the occupation of orbitals by electrons for a particular atom that uses arrows
12. Orbitals may hold no more than two electrons with opposing spins

13. When filling orbitals of equal energy, electrons fill them singly first, with parallel spins
14. Distance the outermost electrons lie from the nucleus of its atom

B. True/False

1. Photons are packets of light energy.
2. The frequency of electromagnetic radiation is proportional to its wavelength.
3. Microwaves have shorter wavelengths than radiowaves.
4. An orbital is a circular path around an atom's nucleus.
5. The $4s$ subshell has higher energy than the $3d$ subshell.
6. Electrons in the same orbital have opposing spins.
7. A carbon atom has six valence electrons.
8. Potassium has a higher ionization energy than sodium.
9. Nitrogen atoms are larger than boron atoms.
10. Lead has more metallic character than tin.

C. Multiple Choice

1. Which of the following types of electromagnetic radiation has the longest wavelength?
    a) Ultraviolet
    b) Visible
    c) Infrared
    d) Microwave

2. Which of the following types of electromagnetic radiation has the highest frequency?
    a) Gamma rays
    b) X-rays
    c) Ultraviolet
    d) Visible

3. How many subshells are in the principal shell $n = 4$?
    a) 1
    b) 2
    c) 3
    d) 4

4. How many orbitals in the $3d$ subshell?
    a) 1
    b) 3
    c) 5
    d) 7

Chapter 9

5. Which element has the electron configuration $1s^2 2s^2 2p^6 3s^2 3p^4$?
   a) C
   b) O
   c) S
   d) Cr

6. Which ion has the electron configuration $1s^2 2s^2 2p^6 3s^2 3p^6 4s^2 3d^{10} 4p^6$?
   a) Ca^{2+}
   b) K^+
   c) Br^-
   d) Kr

7. Which element in the fourth period has six $3d$ electrons?
   a) Cr
   b) Fe
   c) Mo
   d) Ru

8. Which of the following elements has the highest ionization energy?
   a) O
   b) N
   c) P
   d) Al

9. Which of the following atoms is largest?
   a) C
   b) N
   c) O
   d) F

10. Which of the following elements is the most metallic?
    a) Cu
    b) Fe
    c) Cr
    d) Ca

D. Crossword Puzzle

ACROSS

1. Type of spectrum emitted by the atoms of an element after it has absorbed energy
3. Light with wavelengths between those of ultraviolet light and infrared light
4. Type of electrons that are not in the outermost principal shell
5. Distance between adjacent wave crests
6. Fundamental property of electrons represented by the direction of arrows in orbital diagrams
8. Property of elements that increases down a family and from right to left across a period in the periodic table
9. Form of electromagnetic radiation with the shortest wavelengths
10. State of an atom after absorbing energy

DOWN

1. Energy that exhibits both wave-like and particle-light behaviors
2. Number of cycles that pass through a stationary point in one second
3. Type of electrons in the outermost principal shell
7. Packet of light energy

Chapter 9

11. Precise amount
12. Form of electromagnetic radiation with wavelengths between those of visible light and microwaves
13. Type of energy required to remove an electron from the atom in the gaseous state
14. Form of electromagnetic radiation with wavelengths between those of infrared light and radio waves
15. Probability maps that show a statistical distribution of where the electron is likely to be found

ANSWERS TO SELF-TEST QUESTIONS

A. Matching
1. Bohr model 2. quantum-mechanical model 3. electromagnetic spectrum 4. X-rays
5. ultraviolet light 6. radiowaves 7. ground state 8. quantum numbers 9. principal quantum number 10. electron configuration 11. orbital diagram 12. Pauli exclusion principle 13. Hund's rule 14. atomic size

B. True/False
1. T 2. F 3. T 4. F 5. F 6. T 7. F 8. F 9. F 10. T

C. Multiple Choice
1. d 2. a 3. d 4. c 5. c 6. c 7. b 8. a 9. a 10. d

Electrons in Atoms and the Periodic Table

D. Crossword Puzzle

¹E	M	I	S	S	I	O	N				²F		
L											R		
E				³V	I	S	I	B	L	E	E		
⁴C	O	R	E	A							Q		
T				L							U		
R		⁵W	A	V	E	L	E	N	G	T	H		
O				N				⁶S	⁷P	I	N		
⁸M	E	T	A	L	L	I	C		H		C		
A				E					O		Y		
⁹G	A	M	M	A					T				
N									O				
¹⁰E	X	C	I	T	E	D		¹¹Q	U	A	N	T	U	M
T													
¹²I	N	F	R	A	R	E	D		¹³I	O	N	I	Z	A	T	I	O	N

Chapter 9

Chemical Bonding   10

CHAPTER OVERVIEW

Chapter 10 examines Lewis structures for atoms and simple molecules and polyatomic ions. Electronegativity and bond polarity are discussed. Valence shell electron pair repulsion theory is used to predict molecular shape and polarity.

CHAPTER OBJECTIVES

After reading and studying the text, students should be able to:

1. Write Lewis structures for atoms and monoatomic ions.
2. Use Lewis theory to predict the chemical formula of an ionic compound.
3. Differentiate between lone pairs and bonding electron pairs.
4. Describe the differences between single, double, and triple bonds.
5. Write Lewis structures for covalent compounds.
6. Write Lewis structures for polyatomic ions.
7. Write resonance structures for molecules or polyatomic ions with equivalent Lewis structure.
8. Give examples of exceptions to the octet rule.
9. Use valence shell electron pair repulsion theory to predict electron and molecular geometries of simple molecules and polyatomic ions.
10. State the bond angles associated with linear, trigonal planar, and tetrahedral electron pair geometries.
11. Use electronegativities to classify bonds as pure covalent, polar covalent, or ionic.
12. Determine if a molecule is polar or nonpolar.

CHAPTER IN REVIEW

- Bonding theories are models that predict how atoms bond together to form molecules.

- According to Lewis theory, chemical bonds form when atoms transfer or share valence electrons. If the electrons are transferred, the bond is an ionic bond. If the electrons are shared, the bond is a covalent bond.

- Lewis structures show valence electrons as dots surrounding the symbol of an element. The dots are placed around the element's symbol with a maximum of two dots per side.

- Atoms with eight valence electrons, an octet, are particularly stable.

Chapter 10

- Bonding pairs of electrons are electron pairs that are shared between two atoms. They are often represented by dashes in Lewis structures. Lone pairs of electrons are electron pairs that are only on one atom.

- Double bonds consist of two bonding pairs of electrons and triple bonds consist of three pairs of bonding electrons shared between two atoms. Double bonds are shorter and stronger than single bonds. Triple bonds are shorter and stronger than double bonds.

- To write a Lewis structure for a molecular compound, start by writing a skeletal structure. Hydrogen atoms are always terminal. Put less metallic elements in terminal positions and the more metallic elements in central positions. Calculate the total number of valence electrons by summing the valence electrons of each atom in the molecule. Distribute the electrons among the atoms, giving hydrogen a duet and octets to as many atoms as possible. If any atoms lack an octet, move lone pairs of electrons to form double or triple bonds as necessary.

- When writing Lewis structures of polyatomic ions, ionic charge must be taken into account when calculating the number of valence electrons.

- Resonance structures are two or more equivalent, or nearly equivalent, Lewis structures for the same molecule. They are drawn with a double arrow between them. The true structure is an average of the resonance structures.

- Valence shell electron pair repulsion (VSEPR) theory is based on the idea that electron groups—lone pairs, single bonds, and multiple bonds—repel each other. The repulsion between electron groups on the central atom determines the geometry of the molecule.

- To predict geometries using VSEPR theory, draw a Lewis structure for the molecule or polyatomic ion. Count the number of electron groups around the central atom and determine the electron geometry. Count the number of bonding groups and the number of lone pairs around the central atom and determine the molecular geometry.

- Electronegativity is the ability of an element to attract electrons within a covalent bond.

- A dipole moment is a separation of charge within a bond. Covalent bonds that have a dipole moment are called polar covalent bonds.

- A bond formed between two elements of identical electronegativity has no dipole moment. The bond is a nonpolar covalent bond.

- A polar molecule is one with polar bonds that add together—they do not cancel each other—to form a net dipole moment.

Chemical Bonding

SKILLBUILDER PROBLEMS AND SOLUTIONS

SKILLBUILDER 10.1 Writing Lewis Structures for Elements

Write a Lewis structure for Mg.

Solution:
Since Mg is in Group 2A in the periodic table, it has two valence electrons. We represent these as two dots.

Mg:

SKILLBUILDER 10.2 Writing Ionic Lewis Structures

Write a Lewis structure for the compound NaBr.

Solution:
In NaBr, sodium loses its one valence electron, forming a +1 charge, and bromine gains one electron, forming a –1 charge and acquiring an octet.

Na^+ [:Br:]$^-$

SKILLBUILDER 10.3 Using Lewis Theory to Predict the Chemical Formula of an Ionic Compound

Use Lewis theory to predict the formula for the compound that forms between magnesium and nitrogen.

Solution:
Magnesium must lose its two valence electrons (to effectively get an octet in its previous principal shell), while nitrogen needs to gain three electrons to get an octet. Consequently, the compound that forms between Mg and N must have three magnesium atoms to every two nitrogen atoms. The Lewis structure is

Mg^{2+} [:N:]$^{3-}$ Mg^{2+} [:N:]$^{3-}$ Mg^{2+}

The formula is Mg_3N_2.

SKILLBUILDER 10.4 Writing Lewis Structures for Covalent Compounds

Write a Lewis structure for CO.

135

Chapter 10

Solution:
We write

CO

Total number of valence electrons for Lewis structure
  = (# valence e⁻ for C) + (# valence e⁻ for O)
  = 4 + 6 = 10

We place two electrons between the atoms, then we distribute the remaining electrons as lone pairs.

:C–Ö:

The C atom does not have an octet, so we move two lone pairs from the oxygen to form a triple bond with carbon.

:C≡O:

SKILLBUILDER 10.5   **Writing Lewis Structures for Covalent Compounds**

Write a Lewis structure for H_2CO.

Solution:
Following the symmetry guideline, we write

O
H C H

Total number of valence electrons for Lewis structure
  = (# valence e⁻ for C) + (# valence e⁻ for O) + 2(# valence e⁻ for H)

  = 4 + 6 + 2(1) = 12

We distribute the bonding electrons first, then distribute the remaining electrons as lone pairs. To give every atom an octet, we must transfer a lone pair to form a double bond.

Ö:
‖
H—C—H

Chemical Bonding

SKILLBUILDER 10.6 Writing Lewis Structures for Polyatomic Ions

Write a Lewis structure for the ClO⁻ ion.

Solution:
We begin by writing the skeletal structure.

$$Cl \quad O$$

We next calculate the total number of valence electrons for the Lewis structure by summing the number of valence electrons for each atom and adding one electron for the negative charge.

Total number of valence electrons for Lewis structure
= (# valence e⁻ for Cl) + (# valence e⁻ for O) + 1

= 7 + 6 + 1 = 14

We place two electrons between the atoms and then distribute the remaining electrons as lone pairs. Since both atoms have an octet, the Lewis structure is complete. Lastly, we place the Lewis structure in brackets and write the charge of the ion in the upper right-hand corner.

$$[:\!\ddot{\underset{..}{Cl}}\!-\!\ddot{\underset{..}{O}}\!:]^-$$

SKILLBUILDER 10.7 Writing Resonance Structures

Write a Lewis structure for the NO_2^- ion. Include resonance structures.

Solution:
We begin with the skeletal structure. Using the guideline of symmetry, we make the two oxygen atoms terminal.

$$O \quad N \quad O$$

We then calculate the total number of valence electrons for the Lewis structure by summing the number of valence electrons for each atom and adding one electron for the negative charge.

Total number of valence electrons for Lewis structure
= (# valence e⁻ for N) + 2(# valence e⁻ for O) + 1

= 5 + 2(6) + 1 = 18

We next place two electrons between each pair of atoms

$$O-N-O$$

Chapter 10

and then distribute the remaining electrons, first to terminal atoms.

$$:\ddot{\underset{..}{O}} - \ddot{N} - \ddot{\underset{..}{O}}:$$

There are no electrons remaining to complete the octet of the central atom, so we form a double bond by moving a lone pair from one of the oxygen atoms into the bonding region with nitrogen. We also enclose the structure in brackets and include the charge.

$$[\ddot{O} = \ddot{N} - \ddot{\underset{..}{O}}:]^-$$

However, we could have formed the double bond with the other oxygen atom.

$$[:\ddot{\underset{..}{O}} - \ddot{N} = \ddot{O}]^-$$

The two Lewis structures are equally correct, so we write the two structures as resonance structures.

$$[\ddot{O} = \ddot{N} - \ddot{\underset{..}{O}}:]^- \leftrightarrow [:\ddot{\underset{..}{O}} - \ddot{N} = \ddot{O}]^-$$

SKILLBUILDER 10.8   **Predicting Geometry Using VSEPR**

Predict the geometry of ClNO (N is the central atom).

Solution:

$$:\ddot{\underset{..}{Cl}} - \ddot{N} = \ddot{O}$$

ClNO has 18 valence electrons. The central atom (N) has three electron groups (the double bond counts as one group). Two of the three electron groups around N are bonding pairs and one is a lone pair. The electron pair geometry is trigonal planar and the molecular geometry—the shape of the molecule—is bent (3 electron groups, 2 bonding groups, and 1 lone pair).

SKILLBUILDER 10.9   **Predicting Geometry Using VSEPR**

Predict the geometry of the SO_3^{2-} ion.

138

Chemical Bonding

Solution:

$$[:\overset{..}{\underset{..}{O}} - \overset{..}{S} - \overset{..}{\underset{..}{O}}:]^{2-}$$
$$|$$
$$:\overset{..}{\underset{..}{O}}:$$

SO_3^{2-} has 26 valence electrons. The central atom (S) has four electron groups. Three of the four electron groups around S are bonding groups and one is a lone pair. The electron geometry is tetrahedral and the molecular geometry—the shape of the molecule—is trigonal pyramidal.

SKILLBUILDER 10.10 Classifying Bonds as Pure Covalent, Polar Covalent, or Ionic

Determine whether the bond formed between each of the following pairs of atoms is pure covalent, polar covalent, or ionic.

a) I and I
b) Cs and Br
c) P and O

Solution:
a) Since the two atoms are identical, the electronegativity difference is 0. The bond is pure covalent.
b) From Figure 10.2 in the text, we find the electronegativity of Cs (0.7) and of Br (2.8). The electronegativity difference (ΔEN) is

$\Delta EN = 2.8 - 0.7 = 2.1$

Using Table 10.2 in the text, we classify this bond as ionic.
c) From Figure 10.2 in the text, we find the electronegativity of P (2.1) and O (3.5). The electronegativity difference (ΔEN) is

$\Delta EN = 3.5 - 2.1 = 1.4$

Using Table 10.2 in the text, we classify this bond as polar covalent.

SKILLBUILDER 10.11 Determining Whether a Molecule Is Polar

Determine whether CH_4 is polar.

Solution:
From Figure 10.2 in the text, we find the electronegativity of C (2.5) and H (2.1). The electronegativity difference (ΔEN) is 0.4. We classify the bonds as polar covalent, though they are nearly nonpolar. The molecule has tetrahedral electron geometry and tetrahedral molecular

Chapter 10

geometry. The molecule is symmetrical and the small dipole moments of the four identical bonds cancel. There is no net dipole moment. The molecule is nonpolar.

SELF-TEST QUESTIONS

A. Match the following terms with the phrases below.

> bonding pair
> bonding theory
> chemical bond
> Lewis structures
> Lewis theory
> lone pair
> molecular geometry
> octet rule
>
> dipole moment
> electron geometry
> electron group
> polar molecule
> resonance structures
> trigonal planar
> trigonal pyramidal
> valence shell electron pair repulsion theory

1. Model that predicts how atoms bond together to form molecules
2. Bonding theory in which electrons are represented as dots
3. Simple drawings of molecules in which valence electrons surround the symbols of elements
4. Sharing or transfer of electrons to attain stable electron configurations
5. Eight valence electrons gives a stable electron configuration
6. Pair of electrons shared between two atoms
7. Pair of electrons on one atom
8. Two or more equivalent (or nearly equivalent) Lewis structures for the same molecule
9. Theory used to predict the shape of molecules based on electron pair repulsion
10. General term for lone pairs of electrons, single bonds, and multiple bonds
11. Geometric shape with bond angles of 120°
12. Geometric arrangement of electron groups
13. Geometric arrangement of atoms in a molecule
14. Molecular geometry associated with tetrahedral electron pair geometry with three bonding groups
15. Separation of charge within a covalent bond
16. Molecule with a net dipole moment

B. True/False

1. Neon atoms have eight valence electrons.
2. Triple bonds are shorter and stronger than double bonds.
3. All molecules obey the octet rule.
4. Three electron groups around an atom gives a tetrahedral electronic geometry.
5. The bond angle associated with trigonal planar geometry is 180°.
6. Water molecules have a bent geometry.
7. Fluorine is the most electronegative element.
8. Covalent bonds that have a dipole moment are nonpolar bonds.
9. All molecules with polar bonds are polar molecules.

10. Soap molecules interact with both water and grease because they have a polar end and a nonpolar end.

C. Multiple Choice

1. How many valence electrons does a P atom have?
   a) 3
   b) 5
   c) 10
   d) 15

2. Draw the Lewis structure for CO_2. What type of bond is present?
   a) Single
   b) Double
   c) Triple
   d) Ionic

3. How many lone pairs of electrons are in the Lewis structure for N_2?
   a) 1
   b) 2
   c) 3
   d) 4

4. Draw the Lewis structure for the cyanide ion, CN^-. What type of bond is present?
   a) Single
   b) Double
   c) Triple
   d) Ionic

5. The nitrate ion, NO_3^-, is represented by how many resonance structures?
   a) 2
   b) 3
   c) 4
   d) 5

6. Which bond angle is associated with trigonal planar geometry?
   a) 90°
   b) 109.5°
   c) 120°
   d) 180°

7. What is the molecular geometry of CCl_4?
   a) Bent
   b) Trigonal planar
   c) Tetrahedral
   d) Trigonal pyramidal

Chapter 10

8. What is the geometry of the nitrite ion, NO_2^-?
   a) Linear
   b) Bent
   c) Trigonal planar
   d) Tetrahedral

9. Which of the following atoms has the highest electronegativity?
   a) S
   b) Cl
   c) Se
   d) Br

10. Which of the following molecules is polar?
    a) CO_2
    b) CCl_4
    c) NH_3
    d) Br_2

Chemical Bonding

D. Crossword Puzzle

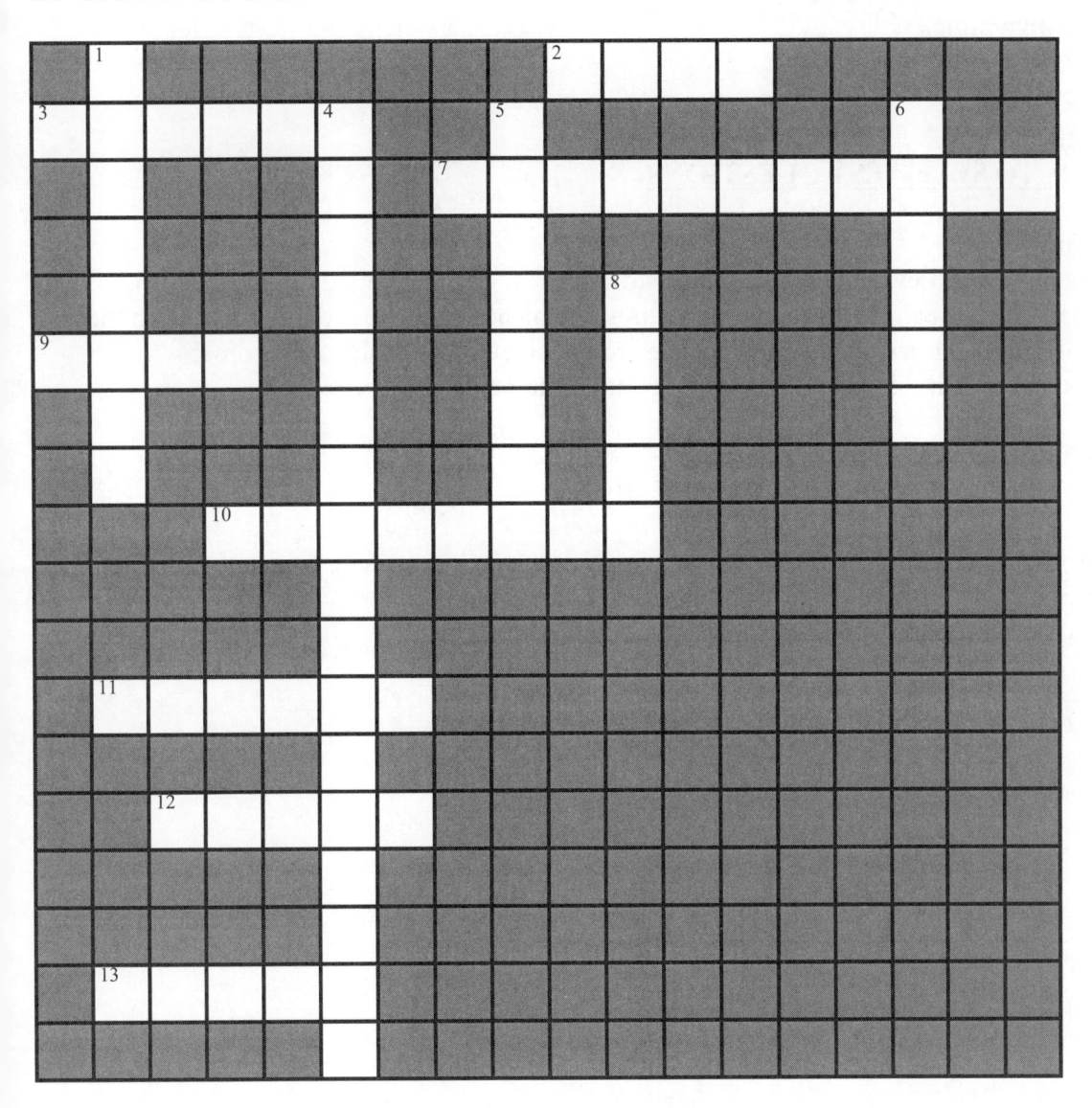

ACROSS

2. Two electrons (one pair)
3. Type of bond in which two electron pairs are shared between two atoms
7. Geometric shape with bond angles of 109.5°
9. Molecular geometry associated with tetrahedral electron pair geometry with two bonding groups
10. No dipole moment
11. Geometric shape with a bond angle of 180°

DOWN

1. Type of bond in which electrons are shared
4. Ability of an element to attract electrons within a covalent bond
5. Type of atom at the end of a molecule
6. Type of bond in which three electron pairs are shared between two atoms
8. Type of covalent bond that has a dipole moment

Chapter 10

12. Type of bond in which electrons are transferred to attain stable electron configurations
13. Eight electrons

ANSWERS TO SELF-TEST QUESTIONS

A. Matching
1. bonding theory 2. Lewis theory 3. Lewis structures 4. chemical bond 5. octet rule 6. bonding pair 7. lone pair 8. resonance structures 9. valence shell electron pair repulsion theory 10. electron group 11. trigonal planar 12. electron geometry 13. molecular geometry 14. trigonal pyramidal 15. dipole moment 16. polar molecule

B. True/False
1. T 2. T 3. F 4. F 5. F 6. T 7. T 8. F 9. F 10. T

C. Multiple Choice
1. b 2. b 3. b 4. c 5. b 6. c 7. c 8. b 9. b 10. c

D. Crossword Puzzle

Across:
2. DUET
3. DOUBLE
7. TETRAHEDRAL
9. BENT
10. NONPOLAR
11. LINEAR
12. IONIC
13. OCTET

Down:
1. COVALENT
4. ELECTRON
5. TERMINAL
6. TRIPLE
7. TRIGONAL
8. POLAR
10. NONPOLAR
12. ELECTRONEGATIVITY

Chapter 10

Gases   11

CHAPTER OVERVIEW

Chapter 11 examines kinetic molecular theory and the properties of gases. Pressure and unit conversions are presented. Gas laws showing relationships among the pressure, volume, temperature, and number of moles of gas are examined. The concept of partial pressure and its application to collecting gases over water is presented. Stoichiometric calculations involving gases are discussed.

CHAPTER OBJECTIVES

After reading and studying the text, students should be able to:

1. State the assumptions of the kinetic molecular theory.
2. Explain how the kinetic molecular theory predicts the properties of gases.
3. Define pressure.
4. Convert between pressure units.
5. State Boyle's law.
6. Use Boyle's law to compute the volume of a gas following a pressure change or the pressure of a gas following a volume change when the temperature and the amount of gas remain constant.
7. State Charles's law.
8. Use Charles's law to compute the volume of gas following a temperature change or the temperature of a gas following a volume change when the pressure and the amount of gas remain constant.
9. State the combined gas law.
10. Use the combined gas law to compute the volume, pressure, or temperature of a gas following a change in the other two conditions when the amount of gas remains constant.
11. State Avogadro's law.
12. Use Avogadro's law to compute the volume of a gas following a change in the amount of gas when the pressure and temperature remain constant.
13. Use the ideal gas law to determine the pressure, temperature, volume, or number of moles of gas, given the other three.
14. Use the ideal gas law to find the molar mass of a gas, given the mass, temperature, pressure, and volume of the gas.
15. State Dalton's law.
16. Calculate the partial pressure of a gas, given the total pressure and the component's fractional composition in a gaseous mixture.
17. State the relationship between vapor pressure and temperature.

Chapter 11

18. Use Dalton's law to compute the partial pressure of a gas collected over water, given the total pressure and the vapor pressure of water.
19. Give the values for standard temperature and pressure (STP) of a gas.
20. Calculate the molar volume of a gas at STP.
21. Carry out stoichiometric calculations for chemical reactions involving gaseous reactants or products.

CHAPTER IN REVIEW

- The newton (N) is a metric unit of force. At sea level, the earth's atmosphere exerts an average pressure of 101,325 N/m^2 or 14.7 lb/in.^{2}.

- Kinetic molecular theory predicts the correct behavior for most gases under many conditions.

- Kinetic molecular theory states that a gas is a collection of particles (atoms or molecules) in constant, straight-line motion. There are no attractions or repulsions between particles. The space between particles is very large compared to the size of the particles themselves. The average kinetic energy of the particles is proportional to the temperature of the gas in kelvins. The higher the temperature, the more energy the particles have and the faster they move.

- Kinetic molecular theory is consistent with gas properties. Gases are compressible, they assume the shape and volume of their container, and they have low densities in comparison with liquids and solids.

- Gas pressure is the force per unit area that results from the constant collisions between the atoms or molecules in the gas and the surrounding surfaces.

- The SI unit of pressure is the pascal (Pa). 1 Pa = 1 N/m^2. Other units of pressure are atmosphere (atm), millimeter of mercury (mm Hg) or torr, pounds per square inch (psi), and inches of mercury (in. Hg).

- Conversions from one pressure unit to another can be accomplished using the relationships in Table 11.1 in the text.

- Boyle's law states that the volume of a gas and its pressure are inversely proportional when the temperature and the amount of gas remain constant.

- Charles's law states that the volume of a gas and its temperature in kelvins are directly proportional when the pressure and the amount of gas remain constant.

- The combined gas law unites Boyle's law and Charles's law into a single relationship.

Gases

- Avogadro's law states that the volume of a gas and the amount of gas in moles are directly proportional when the pressure and temperature of the gas remain constant.

- Gay-Lussac's law states the pressure and temperature of a gas in kelvins are directly proportional when the volume and the amount of gas remain constant.

- The ideal gas law interrelates the four properties of gases—pressure, volume, temperature, and number of moles.

- The ideal gas law can be used in conjunction with mass measurements to calculate the molar mass of a gas.

- The pressure due to an individual component in a gas mixture is called the partial pressure of that component.

- Dalton's law of partial pressures states that the sum of the partial pressures of the components in a gas mixture equals the total pressure.

- When a gas is collected over water, the gas is mixed with water vapor.

- The vapor pressure of water increases with increasing temperature.

- Stoichiometric calculations involving gases are similar to those involving liquids and solids. Coefficients in balanced equations give mole-to-mole relationships. The ideal gas law can be used to convert the volume of a gaseous reactant or product at a specified temperature and pressure to number of moles.

- Conditions for the standard temperature and pressure (STP) of a gas are 273 K and 1 atm.

- The volume occupied by one mole of a gas at STP is called the molar volume and has a value of 22.4 L. The molar volume of a gas can be used to convert between the volume of a gas at STP and the number of moles of the gas.

SKILLBUILDER PROBLEMS AND SOLUTIONS

SKILLBUILDER 11.1   **Converting between Pressure Units**

Convert a pressure of 173 in. Hg into pounds per square inch.

Given: 173 in. Hg

Find: psi

Chapter 11

Conversion Factors:

$$1 \text{ atm} = 29.92 \text{ in. Hg}$$

$$1 \text{ atm} = 14.7 \text{ psi}$$

Solution Map:

$$\boxed{\text{in. Hg}} \rightarrow \boxed{\text{atm}} \rightarrow \boxed{\text{psi}}$$

$$\frac{1 \text{ atm}}{29.92 \text{ in. Hg}} \quad \frac{14.7 \text{ psi}}{1 \text{ atm}}$$

Solution:

$$173 \text{ in. Hg} \times \frac{1 \text{ atm}}{29.92 \text{ in. Hg}} \times \frac{14.7 \text{ psi}}{1 \text{ atm}} = 85.0 \text{ psi}$$

SKILLBUILDER PLUS

Convert a pressure of 23.8 in. Hg into kPa.

Given: 23.8 in. Hg

Find: kPa

Conversion Factors:

29.92 in. Hg = 1 atm

1 atm = 101,325 Pa

1 kPa = 1000 Pa

Solution Map:

$$\boxed{\text{in. Hg}} \rightarrow \boxed{\text{atm}} \rightarrow \boxed{\text{Pa}} \rightarrow \boxed{\text{kPa}}$$

$$\frac{1 \text{ atm}}{29.92 \text{ in. Hg}} \quad \frac{101{,}325 \text{ Pa}}{1 \text{ atm}} \quad \frac{1 \text{ kPa}}{1000 \text{ Pa}}$$

Gases

Solution:

$$23.8 \text{ in. Hg} \times \frac{1 \text{ atm}}{29.92 \text{ in. Hg}} \times \frac{101{,}325 \text{ Pa}}{1 \text{ atm}} \times \frac{1 \text{ kPa}}{1000 \text{ Pa}} = 80.6 \text{ kPa}$$

SKILLBUILDER 11.2   Boyle's Law

A snorkeler takes a syringe filled with 16 mL of air from the surface, where the pressure is 1.0 atm, to an unknown depth. The volume of the air in the syringe at this depth is 7.5 mL. What is the pressure at this depth? If the pressure increases by an additional 1 atm for every 10 m of depth, how deep is the snorkeler?

Given:
$V_1 = 16$ mL
$P_1 = 1.0$ atm
$V_2 = 7.5$ mL

Find: P_2, m

Equation:

$$P_1 V_1 = P_2 V_2$$

Conversion Factor:

$$1 \text{ atm} \equiv 10 \text{ m}$$

Solution Map:

$$\boxed{P_1, V_1, V_2} \rightarrow \boxed{P_2}$$

$$P_1 V_1 = P_2 V_2$$

Solution:

$$P_1 V_1 = P_2 V_2$$

$$P_2 = \frac{P_1}{V_2} V_1$$

$$= \frac{1.0 \text{ atm}}{7.5 \text{ mL}} \; 16 \text{ mL}$$

$$= 2.1 \text{ atm}$$

Chapter 11

The pressure increased from 1.0 atm to 2.1 atm, a change of 1.1 atm. Next we find the depth of the snorkeler.

$$1.1 \text{ atm} \times \frac{10 \text{ m}}{1 \text{ atm}} = 11 \text{ m}$$

SKILLBUILDER 11.3   **Charles's Law**

A gas in a cylinder with a movable piston with an initial volume of 88.2 mL is heated from 35 °C to 155 °C. What is the final volume of the gas (in milliliters)?

Given:
$V_1 = 88.2$ mL
$t_1 = 35$ °C
$t_2 = 155$ °C

Find: V_2

Equation:

$$\frac{V_1}{T_1} = \frac{V_2}{T_2}$$

Solution Map:

$$\boxed{V_1, T_1, T_2} \rightarrow \boxed{V_2}$$

$$\frac{V_1}{T_1} = \frac{V_2}{T_2}$$

Solution:

$$\frac{V_1}{T_1} = \frac{V_2}{T_2}$$

$$V_2 = \frac{V_1}{T_1} T_2$$

Before we substitute in the numerical values, we must convert the temperatures to kelvins (K).

$$T_1 = 35 + 273 = 308 \text{ K}$$

$$T_2 = 155 + 273 = 428 \text{ K}$$

$$V_2 = \frac{V_1}{T_1}T_2$$

$$= \frac{88.2 \text{ mL}}{308 \text{ K}} \cdot 428 \text{ K}$$

$$= 123 \text{ mL}$$

SKILLBUILDER 11.4   The Combined Gas Law

A balloon has a volume of 3.7 L at a pressure of 1.1 atm and a temperature of 30 °C. If the balloon is submerged in water to a depth where the pressure is 4.7 atm and the temperature is 15 °C, what will its volume be (assume that any changes in pressure caused by the skin of the balloon are negligible)?

Given:
- $V_1 = 3.7$ L
- $P_1 = 1.1$ atm
- $t_1 = 30$ °C
- $P_2 = 4.7$ atm
- $t_2 = 15$ °C

Find: V_2

Equation:

$$\frac{P_1 V_1}{T_1} = \frac{P_2 V_2}{T_2}$$

Solution Map:

$$\boxed{P_1, V_1, T_1, P_2, T_2} \rightarrow \boxed{V_2}$$

$$\frac{P_1 V_1}{T_1} = \frac{P_2 V_2}{T_2}$$

Solution:

$$\frac{P_1 V_1}{T_1} = \frac{P_2 V_2}{T_2}$$

$$V_2 = \frac{P_1 V_1 T_2}{T_1 P_2}$$

Before substituting into the equation, we must convert the temperatures to kelvins.

Chapter 11

$$T_1 = 30 + 273 = 303 \text{ K}$$
$$T_2 = 15 + 273 = 288 \text{ K}$$

Now we can substitute into the equation and compute V_2.

$$P_2 = \frac{1.1 \text{ atm} \times 3.7 \text{ L} \times 288 \text{ K}}{303 \text{ K} \times 4.7 \text{ atm}}$$

$$= 0.82 \text{ L}$$

SKILLBUILDER 11.5   Avogadro's Law

A chemical reaction occurring at constant temperature in a cylinder equipped with a movable piston produces 0.58 mol of a gaseous product. If the cylinder contained 0.11 mol of gas before the reaction and had an initial volume of 2.1 L, what was its volume after the reaction? (All of the gas initially present reacted.)

Given:
$n_1 = 0.11$ mol
$V_1 = 2.1$ L
$n_2 = 0.58$ mol

Find: V_2

Equation:

$$\frac{V_1}{n_1} = \frac{V_2}{n_2}$$

Solution Map:

$$\boxed{n_1, V_1, n_2} \rightarrow \boxed{V_2}$$

$$\frac{V_1}{n_1} = \frac{V_2}{n_2}$$

Solution:

$$\frac{V_1}{n_1} = \frac{V_2}{n_2}$$

$$V_2 = \frac{n_2}{n_1} V_1$$

$$V_2 = \frac{0.58 \text{ mol}}{0.11 \text{ mol}} \; 2.1 \text{ L}$$

$$= 11 \text{ L}$$

SKILLBUILDER 11.6   **The Ideal Gas Law**

An 8.5-L tire is filled with 0.55 mol of gas at a temperature of 305 K. What is the pressure of the gas in the tire?

Given:
    $V = 8.5$ L
    $n = 0.55$ mol
    $T = 305$ K

Find: P

Equation:

$$PV = nRT$$

Solution Map:

$$\boxed{P, n, T} \;\;\rightarrow\;\; \boxed{V}$$

$$PV = nRT$$

Solution:

$$PV = nRT$$

$$P = \frac{nRT}{V}$$

$$= \frac{0.55 \text{ mol} \times 0.0821 \frac{\text{L} \cdot \text{atm}}{\text{K} \cdot \text{mol}} \times 305 \text{ K}}{8.5 \text{ L}}$$

$$= 1.6 \text{ atm}$$

Chapter 11

SKILLBUILDER 11.7 **The Ideal Gas Law Requiring Unit Conversion**

How much volume does 0.556 mol of gas occupy when its pressure is 715 mm Hg and its temperature is 58 °C?

Given:
$n = 0.556$ mol
$P = 715$ mm Hg
$t = 58$ °C

Find: V

Equation:

$$PV = nRT$$

Solution Map:

$$\boxed{P, n, T} \rightarrow \boxed{V}$$

$$PV = nRT$$

Solution:

$$PV = nRT$$

$$V = \frac{nRT}{P}$$

Before substituting into the equation, we must convert P and t into the correct units.

$$P = 715 \text{ mm Hg} \times \frac{1 \text{ atm}}{760 \text{ mm Hg}} = 0.94\underline{0}78 \text{ atm}$$

$$T = 58 + 273 = 331 \text{ K}$$

Now we can substitute into the equation and compute V.

$$V = \frac{0.556 \text{ mol} \times 0.0821 \frac{\text{L} \cdot \text{atm}}{\text{K} \cdot \text{mol}} \times 331 \text{ K}}{0.94\underline{0}78 \text{ atm}}$$

$$= 16.1 \text{ L}$$

Gases

SKILLBUILDER PLUS

Find the pressure in millimeters of mercury of a 0.133-g sample of helium gas at 32 °C and contained in a 648-mL container.

Given:
$m = 0.133$ g
$t = 32$ °C
$V = 648$ mL

Find: P

Equation and Conversion Factors:

$PV = nRT$
1 mol He = 4.00 g He
760 mm Hg = 1 atm

Solution Maps:

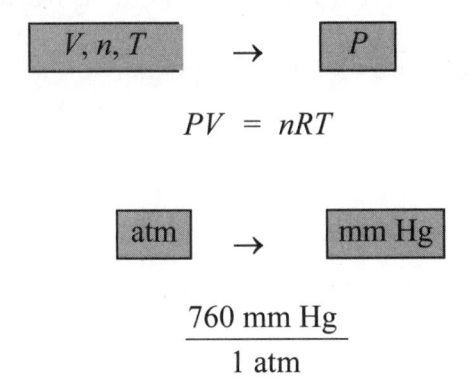

$$\boxed{V, n, T} \rightarrow \boxed{P}$$

$$PV = nRT$$

$$\boxed{\text{atm}} \rightarrow \boxed{\text{mm Hg}}$$

$$\frac{760 \text{ mm Hg}}{1 \text{ atm}}$$

Solution:

$$PV = nRT$$

$$P = \frac{nRT}{V}$$

Before substituting into the equation, we must convert g to mol and convert V and t into the correct units.

$$n = 0.133 \text{ g} \times \frac{1 \text{ mol He}}{4.00 \text{ g He}} = 0.0332\underline{5} \text{ mol He}$$

Chapter 11

$$V = 648 \text{ mL} \times \frac{0.001 \text{ L}}{1 \text{ mL}} = 0.648 \text{ L}$$

$$T = 32 + 273 = 305 \text{ K}$$

Now we can substitute into the equation and compute P.

$$P = \frac{0.033\underline{2}5 \text{ mol} \times 0.0821 \frac{\text{L} \cdot \text{atm}}{\text{K} \cdot \text{mol}} \times 305 \text{ K}}{0.648 \text{ L}}$$

$$= 1.28 \text{ atm}$$

Lastly, we convert the pressure unit to mm Hg.

$$1.28 \text{ atm} \times \frac{760 \text{ mm Hg}}{1 \text{ atm}} = 973 \text{ mm Hg}$$

SKILLBUILDER 11.8   Molar Mass Using the Ideal Gas Law and Mass Measurement

A sample of gas has a mass of 827 mg. Its volume is 0.270 L at a temperature of 88 °C and a pressure of 975 mm Hg. Find its molar mass.

Given:
  $m = 827$ mg
  $V = 0.270$ L
  $t = 88$ °C
  $P = 975$ mm Hg

Find: molar mass (g/mol)

Equations:

$$PV = nRT$$

$$\text{Molar mass} = \frac{\text{Mass } (m)}{\text{Moles } (n)}$$

Solution Map:

$$\boxed{P, V, T} \rightarrow \boxed{n}$$

$$PV = nRT$$

158

$n, m \rightarrow$ Molar mass

$$\text{Molar mass} = \frac{\text{Mass } (m)}{\text{Moles } (n)}$$

Solution:

$$PV = nRT$$

$$n = \frac{PV}{RT}$$

Before we substitute into the equation, we must convert the pressure to atm and temperature to K.

$$P = 975 \text{ mm Hg} \times \frac{1 \text{ atm}}{760 \text{ mm Hg}} = 1.2\underline{8}28 \text{ atm}$$

$$T = 88 + 273 = 361 \text{ K}$$

Now we can substitute into the equation and compute n, the number of moles.

$$n = \frac{1.2\underline{8}28 \text{ atm} \times 0.270 \text{ L}}{0.0821 \frac{\text{L} \cdot \text{atm}}{\text{K} \cdot \text{mol}} \times 361 \text{ K}}$$

$$= 0.011\underline{6}86 \text{ mol}$$

Before we substitute into the equation to compute the molar mass, we must convert the mass into the correct unit.

$$827 \text{ mg} \times \frac{0.001 \text{ g}}{1 \text{ mg}} = 0.827 \text{ g}$$

Finally, we use the number of moles (n) and the mass (m) to find the molar mass.

$$\text{Molar mass} = \frac{\text{Mass } (m)}{\text{Moles } (n)}$$

$$= \frac{0.827 \text{ g}}{0.011\underline{6}86 \text{ mol}}$$

$$= 70.8 \text{ g/mol}$$

Chapter 11

SKILLBUILDER 11.9   Total Pressure and Partial Pressure

A sample of hydrogen gas is mixed with water vapor. The mixture has a total pressure of 745 torr and the water vapor has a partial pressure of 24 torr. What is the partial pressure of the hydrogen gas?

Given:
$$P_{tot} = 745 \text{ torr}$$
$$P_{H_2O} = 24 \text{ torr}$$

Find: P_{H_2}

Equation:
$$P_{tot} = P_a + P_b + P_c + ...$$

Solution:
$$P_{tot} = P_{H_2} + P_{H_2O}$$
$$P_{H_2} = P_{tot} - P_{H_2O}$$
$$= 745 \text{ torr} - 24 \text{ torr}$$
$$= 721 \text{ torr}$$

SKILLBUILDER 11.10   Partial Pressure, Total Pressure, and Percent Composition

What must the total pressure be for a diver breathing heliox with an oxygen composition of 5.0% to breath $P_{O_2} = 0.21$ atm?

Given:
Composition O_2 = 5.0%
$P_{O_2} = 0.21$ atm

Find: P_{tot}

Equation:

Partial pressure of component =
    Fractional composition of component × Total pressure

Solution:
Total pressure =
    Partial pressure of component ÷ Fractional composition of component

Gases

The fractional composition of O₂ is simply the percent composition divided by 100.

$$\text{Fractional composition of } O_2 = \frac{5.0}{100} = 0.050$$

$$P_{tot} = \frac{0.21 \text{ atm}}{0.050}$$

$$= 4.2 \text{ atm}$$

SKILLBUILDER 11.11 — Gases in Chemical Reactions

In the following reaction, 4.58 L of O₂ were formed at 745 mm Hg and 308 K. How many grams of Ag₂O must have decomposed?

$$2 \text{ Ag}_2O(s) \rightarrow 4 \text{ Ag}(s) + O_2(g)$$

Given:
- $V = 4.58$ L
- $P = 745$ mm Hg
- $T = 308$ K

Find: g Ag₂O

Equation and Conversion Factors:

$PV = nRT$

2 mol Ag₂O ≡ 1 mol O₂

1 mol Ag₂O = 231.74 g Ag₂O

Solution Map:

P, V, T → n (mol O₂)

$PV = nRT$

mol O₂ → mol Ag₂O → g Ag₂O

$\dfrac{2 \text{ mol Ag}_2O}{1 \text{ mol O}_2}$ $\dfrac{231.74 \text{ g Ag}_2O}{1 \text{ mole Ag}_2O}$

Solution:

$PV = nRT$

Chapter 11

$$n = \frac{PV}{RT}$$

Before we substitute the values into this equation, we must convert the pressure to atm.

$$P = 745 \text{ mm Hg} \times \frac{1 \text{ atm}}{760 \text{ mm Hg}} = 0.98\underline{0}26 \text{ atm}$$

We now substitute the appropriate values and compute n.

$$n = \frac{PV}{RT}$$

$$n = \frac{0.98\underline{0}26 \text{ atm} \times 4.58 \text{ L}}{0.0821 \frac{\text{L} \cdot \text{atm}}{\text{K} \cdot \text{mol}} \times 308 \text{ K}}$$

$$= 0.17\underline{7}54 \text{ mol O}_2$$

$$0.17\underline{7}54 \text{ mol O}_2 \times \frac{2 \text{ mol Ag}_2\text{O}}{1 \text{ mol O}_2} \times \frac{231.74 \text{ g O}_2}{1 \text{ mol Ag}_2\text{O}} = 82.3 \text{ g Ag}_2\text{O}$$

SKILLBUILDER 11.12 Using Molar Volume in Calculations

How many liters of oxygen (at STP) are required to form 10.5 g of H_2O?

$$2 \text{ H}_2(g) + \text{O}_2(g) \rightarrow 2 \text{ H}_2\text{O}(g)$$

Given: 10.5 g H_2O

Find: L O_2

Conversion Factors:

1 mol H_2O = 18.02 g H_2O
2 mol H_2O ≡ 1 mol O_2
1 mol O_2 = 22.4 L O_2 (at STP)

Solution Map:

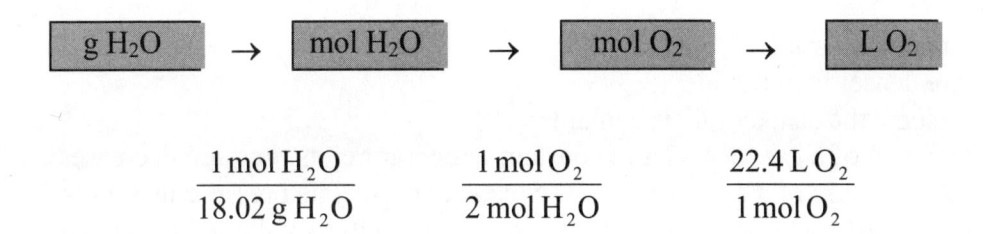

g H₂O → mol H₂O → mol O₂ → L O₂

$$\frac{1 \text{ mol H}_2\text{O}}{18.02 \text{ g H}_2\text{O}} \qquad \frac{1 \text{ mol O}_2}{2 \text{ mol H}_2\text{O}} \qquad \frac{22.4 \text{ L O}_2}{1 \text{ mol O}_2}$$

Solution:

$$10.5 \text{ g H}_2\text{O} \times \frac{1 \text{ mol H}_2\text{O}}{18.02 \text{ g H}_2\text{O}} \times \frac{1 \text{ mol O}_2}{2 \text{ mol H}_2\text{O}} \times \frac{22.4 \text{ L O}_2}{1 \text{ mol O}_2} = 6.53 \text{ L O}_2$$

SELF-TEST QUESTIONS

A. Match the following terms with the phrases below.

absolute zero
Avogadro's law
Boyle's law
Charles's law
combined gas law
Dalton's law of partial pressures
ideal gas law
ideal gas constant

kinetic molecular theory
millimeter of mercury (mm Hg)
molar volume
nitrogen narcosis
oxygen toxicity
partial pressure
vapor pressure

1. Model that predicts the behavior of gases
2. Unit of pressure associated with the way in which pressure is measured with a barometer
3. The volume of a gas and its pressure are inversely proportional
4. 0 K
5. The volume of a gas and its kelvin temperature are directly proportional
6. Gas law that gives the relationship between the temperature, pressure, and volume of a gas
7. The volume of a gas and the amount of the gas in moles are directly proportional
8. Constant represented by R in the ideal gas equation
9. Gas law that relates the volume, temperature, pressure, and number of moles of a gas
10. Pressure due to an individual component in a gas mixture
11. The sum of the partial pressures of each of the components in a gas mixture equals the total pressure
12. Partial pressure of gas phase water in contact with liquid water
13. Physiological condition resulting from high oxygen concentrations in bodily tissues
14. Physiological condition resulting from high nitrogen concentrations in bodily tissues and fluids
15. Volume occupied one mole of a gas at standard temperature and pressure

Chapter 11

B. True/False

1. Gases are compressible.
2. Gases are denser than liquids.
3. Gases assume the shape of their container.
4. As the volume of a gas increases at constant temperature, its pressure decreases.
5. As the temperature of a gas increases at constant volume, its pressure increases.
6. If the temperature and pressure of a gas remain constant, its volume increases as the number of moles of gas increases.
7. The pressure exerted by a particular gas in a mixture is independent of the presence of the other gases in the mixture.
8. Oxygen toxicity is a condition caused by low oxygen levels in bodily tissues.
9. Gases collected over water are mixed with water vapor.
10. The molar volume of a gas is 22.4 L under all conditions of temperature and pressure.

C. Multiple Choice

1. Convert 752 torr to psi.
   a) 0.0673 psi
   b) 0.989 psi
   c) 14.5 psi
   d) 51.2 psi

2. A sample of gas has an initial volume of 3.62 L at a pressure of 0.987 atm. If the volume of the gas decreases to 2.50 L, what will the pressure be? The temperature does not change.
   a) 0.682 atm
   b) 0.691 atm
   c) 1.43 atm
   d) 3.57 atm

3. A balloon with an initial volume of 1.24 L at a temperature of 24 °C is warmed to 58 °C. What is the volume of the balloon at 58 °C? The pressure does not change.
   a) 0.513 L
   b) 1.11 L
   c) 1.38 L
   d) 3.00 L

4. A balloon contains 0.148 mol of helium and has a volume of 3.06 L. If helium is added and the balloon expands to 4.24 L, how many moles of helium were added? The pressure and temperature remain constant.
   a) 0.057 mol
   b) 0.107 mol
   c) 0.205 mol
   d) 0.353 mol

Gases

5. A gas sample occupies 255 mL at 26 °C and a pressure of 748 torr. The temperature decreases to 18 °C and the pressure drops to 739 torr. What is the final volume of the gas?
   a) 179 mL
   b) 251 mL
   c) 259 mL
   d) 265 mL

6. What is the volume of 0.145 mol of gas at 68 °C, 1.12 atm?
   a) 0.723 L
   b) 0.907 L
   c) 3.62 L
   d) 11.0 L

7. A gaseous mixture of N_2, Ne, and O_2 has a total pressure of 836 torr. The partial pressure of N_2 is 134 torr and the partial pressure of Ne is 204 torr. What is the partial pressure of O_2?
   a) 338 torr
   b) 498 torr
   c) 702 torr
   d) 1174 torr

8. What is the volume of 0.256 mol O_2 gas at STP?
   a) 5.73 mL
   b) 87.5 mL
   c) 184 mL
   d) 5.73 L

9. A sample of gas in a 225 mL vessel has a mass of 0.749 g. The pressure is 756 torr and the temperature is 32 °C. What is the molar mass of the gas?
   a) 18.9 g/mol
   b) 82.9 g/mol
   c) 83.8 g/mol
   d) 112 g/mol

10. The hydrogen gas produced by the reaction of magnesium with hydrochloric acid is collected over water at 25 °C and a total pressure of 751 torr. The volume of the gas is 28.8 mL. The vapor pressure of water at 22 °C is 23.8 torr. How many grams of magnesium reacted?

    $$Mg(s) + 2\,HCl(aq) \rightarrow MgCl_2(aq) + H_2(g)$$

    a) 0.0274 g
    b) 0.0283 g
    c) 0.0286 g
    d) 0.326 g

Chapter 11

D. Crossword Puzzle

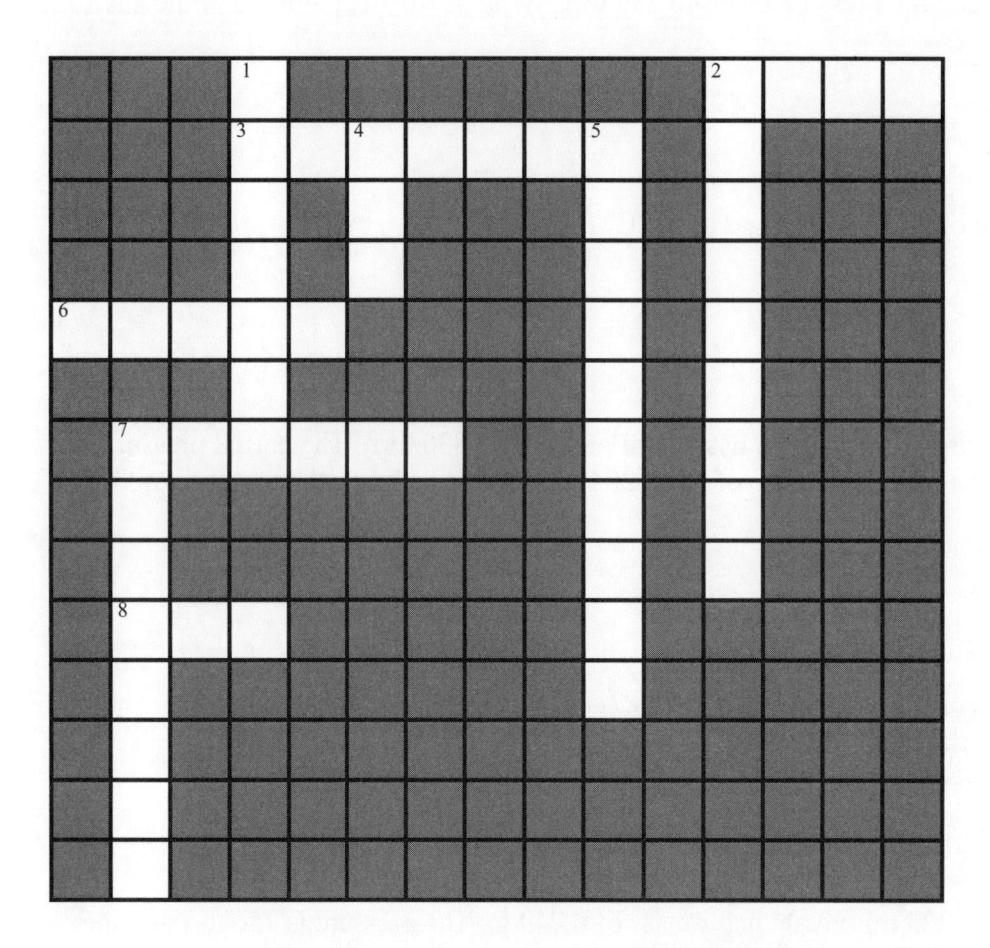

ACROSS

2. Unit of pressure named after the inventor of the barometer
3. Physiological condition resulting from low oxygen levels
6. Scientist who discovered the relationship between gas volume and pressure
7. Unit of pressure symbolized as Pa
8. 273 K, 1 atm

DOWN

1. French mathematician and physicist who quantified the relationship between the volume of a gas and its temperature
2. Italian physicist who invented the barometer
4. Pounds per square inch
5. Pressure unit symbolized as atm
7. Force per unit area

Gases

ANSWERS TO SELF-TEST QUESTIONS

A. Matching
1. kinetic molecular theory 2. millimeter of mercury (mm Hg) 3. Boyle's law 4. absolute zero
5. Charles's law 6. combined gas law 7. Avogadro's law 8. ideal gas constant (*R*) 9. ideal gas law 10. partial pressure 11. Dalton's law of partial pressures 12. vapor pressure 13 oxygen toxicity 14. nitrogen narcosis 15. molar volume

B. True/False
1. T 2. F 3. T 4. T 5. T 6. T 7. T 8. F 9. T 10. F

C. Multiple Choice
1. c 2. c 3. c 4. a 5. b 6. c 7. b 8. d 9. c 10. a

D. Crossword Puzzle

			¹C						²T	O	R	R
			³H	Y	⁴P	O	X	I	⁵A		O		
			A		S				T		R		
			R		I				M		I		
⁶B	O	Y	L	E					O		C		
			E						S		E		
	⁷P	A	S	C	A	L			P		L		
	R								H		L		
	E								E		I		
	⁸S	T	P						R				
	S								E				
	U												
	R												
	E												

167

Chapter 11

Liquids, Solids, and Intermolecular Forces

12

CHAPTER OVERVIEW

Chapter 12 focuses on interactions between molecules. Intermolecular forces and their influence on the properties of liquids and solids are described.

CHAPTER OBJECTIVES

After reading and studying the text, students should be able to:

1. Describe the properties of liquids and solids.
2. Explain how the strength of intermolecular forces affect the surface tension and viscosity of a liquid.
3. List the factors that affect the rate of vaporization of a liquid.
4. List the factors that affect the vapor pressure of a liquid.
5. Define boiling point and normal boiling point.
6. Define heat of vaporization.
7. Use the heat of vaporization to calculate the amount of heat energy required to vaporize a given amount of a liquid.
8. Define heat of fusion.
9. Use the heat of fusion to calculate the amount of heat energy required to melt a given amount of a solid.
10. Describe the processes of evaporation, condensation, melting, freezing, and sublimation from a molecular viewpoint.
11. Characterize the processes of evaporation, condensation, melting, freezing, and sublimation as endothermic or exothermic.
12. Describe the three different types of intermolecular forces and their effects on the physical properties of liquids.
13. Use intermolecular forces to determine relative melting points and boiling points of compounds.
14. Describe the differences between crystalline solids and amorphous solids.
15. Describe the three types of crystalline solids.
16. Identify solids as molecular, ionic, or atomic.
17. Describe the unique properties of water.

Chapter 12

CHAPTER IN REVIEW

- Liquids have high densities in comparison to gases. They are not easily compressed. Liquids have definite volume but indefinite shape; they assume the shape of their containers.

- Solids have high densities in comparison to gases. They are not easily compressed. Solids have definite volume and definite shape. Solids may be crystalline (ordered) or amorphous (disordered).

- Surface tension is the tendency of liquids to minimize their surface area.

- Viscosity is the resistance of a liquid to flow.

- Evaporation, or vaporization, is an endothermic physical change in which a substance converts from its liquid form to its gas form.

- The rate of vaporization of a liquid increases with increasing surface area, increasing temperature, and decreasing strength of intermolecular forces.

- Liquids that evaporate easily are volatile. Liquids that do not evaporate easily are nonvolatile.

- Condensation is an exothermic physical change in which a substance converts from its gas form to its liquid form.

- Vapor pressure is the partial pressure of a gas in dynamic equilibrium with its liquid.

- Vapor pressure increases with increasing temperature and decreasing strength of intermolecular forces.

- The boiling point of a substance is the temperature at which the vapor pressure of its liquid is equal to the pressure above it. The normal boiling point is the temperature at which the vapor pressure of a liquid equals 1 atm.

- Heat of vaporization (ΔH_{vap}) is the amount of heat required to vaporize one mole of a liquid. Heat of vaporization is temperature dependent. The higher the temperature the easier it is to vaporize a liquid and the lower the heat of vaporization.

- Heat of vaporization can be used to calculate the amount of heat energy required to vaporize a given amount of liquid.

- At the melting point of a substance, the atoms and molecules have enough thermal energy to overcome the intermolecular forces that hold them at their stationary points in the solid. The solid converts to a liquid.

Liquids, Solids, and Intermolecular Forces

- Melting is endothermic; freezing is exothermic.

- Heat of fusion (ΔH_{fus}) is the amount of heat required to melt one mole of a solid.

- Heat of fusion can be used to calculate the amount of heat energy required to melt a given amount of solid.

- Sublimation is an endothermic physical change in which a substance is converted from its solid form directly into its gaseous form.

- The strength of the intermolecular forces between the atoms or molecules that compose a substance determines its physical state at room temperature. Intermolecular forces are stronger in solids and liquids than in gases.

- Dispersion forces are intermolecular forces present in all atoms and molecules. They are caused by fluctuations in the electron distribution within molecules or atoms that result in instantaneous or temporary dipoles. An instantaneous dipole in one atom or molecule induces an instantaneous dipole in a neighboring atom or molecule. The partially positive end of one instantaneous dipole is attracted to the partially negative end of another instantaneous dipole.

- Within a family of similar elements or compounds, the magnitude of the dispersion force increases with increasing molar mass.

- Dipole-dipole forces exist in all polar molecules. Polar molecules have permanent dipoles. The partially positive end of one polar molecule is attracted to the partially negative end of another polar molecule.

- Miscibility is the ability of liquids to mix without separating into different phases. Polar liquids are miscible with polar liquids but are not miscible with nonpolar liquids.

- Hydrogen bonding is an intermolecular force present in polar molecules in which hydrogen atoms are bonded directly to fluorine, oxygen, or nitrogen. The hydrogen bond is an attraction between the hydrogen atom and a fluorine, oxygen, or nitrogen atom in a neighboring molecule. Hydrogen bonds are the strongest of the three intermolecular forces.

- Molecular solids are solids whose composite units are molecules.

- Ionic solids are solids whose composite units are formula units.

- Atomic solids are solids whose composite units are individual atoms.

- Atomic solids can be categorized as covalent atomic solids, nonbonding atomic solids, and metallic atomic solids. Covalent atomic solids are held together by covalent bonds. Nonbonding atomic solids are held together by relatively weak dispersion forces.

Chapter 12

Metallic atomic solids are held together by metallic bonds that consist of positively charged metal ions in a sea of electrons.

- Water is unique among liquids. It has a low molar mass and a relatively high boiling point.

- Water molecules have a high dipole moment resulting from polar O-H bonds and a bent molecular geometry. Water molecules form strong hydrogen bonds with other molecules.

- Water expands upon freezing. Ice is less dense than liquid water and floats on the liquid.

SKILLBUILDER PROBLEMS AND SOLUTIONS

SKILLBUILDER 12.1 — Using the Heat of Vaporization in Calculations

Calculate the amount of heat in kilojoules required to vaporize 2.58 kg of water at its boiling point.

Given: 2.58 kg H_2O

Find: kJ

Conversion Factors:

ΔH_{vap} = 40.6 kJ/mol (at 100 °C)

1 mol H_2O = 18.02 g H_2O

1 kg = 1000 g

Solution Map:

kg H_2O → g H_2O → mol H_2O → kJ

$$\frac{1000 \text{ g } H_2O}{1 \text{ kg } H_2O} \quad \frac{1 \text{ mol } H_2O}{18.02 \text{ g } H_2O} \quad \frac{40.6 \text{ kJ}}{1 \text{ mol } H_2O}$$

Solution:

$$2.58 \text{ kg } H_2O \times \frac{1000 \text{ g } H_2O}{1 \text{ kg } H_2O} \times \frac{1 \text{ mol } H_2O}{18.02 \text{ g } H_2O} \times \frac{40.6 \text{ kJ}}{\text{mol } H_2O} = 5.81 \times 10^3 \text{ kJ}$$

Liquids, Solids, and Intermolecular Forces

SKILLBUILDER PLUS

A drop of water weighing 0.48 g condenses on the surface of a 55-g block of aluminum that is initially at 25 °C. If the heat released during condensation goes only toward heating the metal, what is the final temperature in Celsius of the metal block? (The specific heat capacity of aluminum is 0.903 J/g °C.)

Given:
- 0.48 g H$_2$O
- 55 g Al
- $T_i = 25$ °C

Find: T_f

Equations and Conversion Factors:

$q = m \cdot C \cdot \Delta T$

$\Delta T = T_f - T_i$

$C = 0.903$ J/g °C

1 kJ = 1000 J

$\Delta H_{vap} = 40.6$ kJ/mol

1 mol H$_2$O = 18.02 g H$_2$O

Solution Map:

g H$_2$O → mol H$_2$O → kJ

$\dfrac{1 \text{ mol H}_2\text{O}}{18.02 \text{ g H}_2\text{O}} \quad \dfrac{40.6 \text{ kJ}}{1 \text{ mol H}_2\text{O}}$

q, m, C → ΔT

$q = m \cdot C \cdot \Delta T$

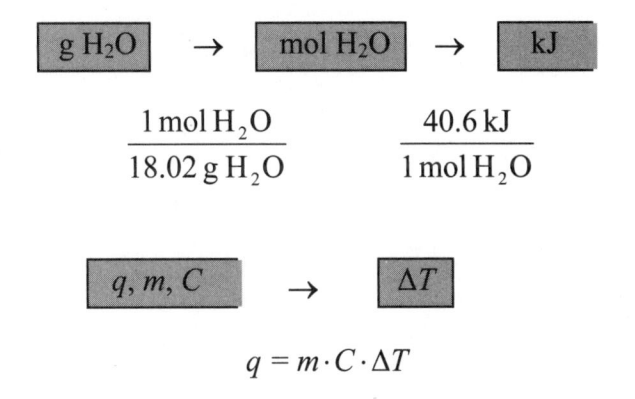

Solution:

$0.48 \text{ g H}_2\text{O} \times \dfrac{1 \text{ mol H}_2\text{O}}{18.02 \text{ g H}_2\text{O}} \times \dfrac{40.6 \text{ kJ}}{1 \text{ mol H}_2\text{O}} = 1.0814 \text{ kJ}$

Chapter 12

The heat released by the condensation of water is equal to the heat absorbed by the aluminum. We use this amount of heat to calculate the temperature change of aluminum.

$$q = m \cdot C \cdot \Delta T$$

$$\Delta T = \frac{q}{m \cdot C}$$

Before we substitute value into the equation, we must convert kJ to J.

$$1.0814 \text{ kJ} \times \frac{1000 \text{ J}}{1 \text{ kJ}} = 1.0814 \times 10^3 \text{ J}$$

$$\Delta T = \frac{1.0814 \times 10^3 \text{ J}}{55 \text{ g} \times 0.903 \frac{\text{J}}{\text{g °C}}} = 22 \text{ °C}$$

Solve the second equation for T_f.

$$\Delta T = T_f - T_i$$

$$T_f = \Delta T + T_i$$

Substitute the correct values and compute the answer.

$$T_f = 22 \text{ °C} + 25 \text{ °C} = 47 \text{ °C}$$

SKILLBUILDER 12.2   Using the Heat of Fusion in Calculations

Calculate the amount of heat absorbed when a 15.5-g ice cube melts.

Given: 15.5 g H_2O

Find: kJ

Conversion Factors:

$$\Delta H_{\text{fus}} = 6.02 \text{ kJ/mol (at 0.00 °C)}$$

$$1 \text{ mol } H_2O = 18.02 \text{ g } H_2O$$

Liquids, Solids, and Intermolecular Forces

Solution Map:

g H₂O → mol H₂O → kJ

$$\frac{1 \text{ mol H}_2\text{O}}{18.02 \text{ g H}_2\text{O}} \qquad \frac{6.02 \text{ kJ}}{1 \text{ mol H}_2\text{O}}$$

Solution:

$$15.5 \text{ g H}_2\text{O} \times \frac{1 \text{ mol H}_2\text{O}}{18.02 \text{ g H}_2\text{O}} \times \frac{6.02 \text{ kJ}}{1 \text{ mol H}_2\text{O}} = 5.18 \text{ kJ}$$

SKILLBUILDER PLUS

A 5.6 g ice cube is placed into 195 g of water initially at room temperature. If the heat absorbed for melting the ice comes only from the 195 g of water, what is the temperature change of the 195 g of water?

Given:
 5.6 g H₂O(s) (ice)
 195 g H₂O(l)

Find: ΔT for H₂O(l)

Equation and Conversion Factors:

$q = m \cdot C \cdot \Delta T$

$C = 4.18 \text{ J/g °C}$

1 kJ = 1000 J

ΔH_fus = 6.02 kJ/mol (at 0.00 °C)

1 mol H₂O = 18.02 g H₂O

Solution Map:

g H₂O → mol H₂O → kJ

$$\frac{1 \text{ mol H}_2\text{O}}{18.02 \text{ g H}_2\text{O}} \qquad \frac{6.02 \text{ kJ}}{1 \text{ mol H}_2\text{O}}$$

Chapter 12

$$\boxed{q, m, C} \rightarrow \boxed{\Delta T}$$

$$q = m \cdot C \cdot \Delta T$$

Solution:

$$5.6 \text{ g H}_2\text{O} \times \frac{1 \text{ mol H}_2\text{O}}{18.02 \text{ g H}_2\text{O}} \times \frac{6.02 \text{ kJ}}{1 \text{ mol H}_2\text{O}} = 1.\underline{8}70 \text{ kJ}$$

The heat absorbed by the melting of ice is equal to the heat released by the liquid water. Since the liquid water releases heat, the value of q is negative. We use this amount of heat to calculate the temperature change.

$$q = m \cdot C \cdot \Delta T$$

$$\Delta T = \frac{q}{m \cdot C}$$

Before we substitute value into the equation, we must convert kJ to J.

$$-1.\underline{8}70 \text{ kJ} \times \frac{1000 \text{ J}}{1 \text{ kJ}} = -1.\underline{8}70 \times 10^3 \text{ J}$$

$$\Delta T = \frac{-1.\underline{8}70 \times 10^3 \text{ J}}{195 \text{ g} \times 4.18 \frac{\text{J}}{\text{g °C}}} = -2.3 \text{ °C}$$

SKILLBUILDER 12.3 Dispersion Forces

Which hydrocarbon, CH_4 or C_2H_6, has the higher boiling point?

Solution:
The molar mass of CH_4 is 16.04 g/mol and the molar mass of C_2H_6 is 30.08 g/mol. Since C_2H_6 has the higher molar mass, it has stronger dispersion forces and therefore a higher boiling point.

SKILLBUILDER 12.4 Dipole-Dipole Forces

Which of the following molecules have dipole-dipole forces?

a) CI_4
b) CH_3Cl
c) HCl

Liquids, Solids, and Intermolecular Forces

Solution:
A molecule will have dipole-dipole forces if it is polar.

a) Since the electronegativities of C and I are both 2.5 (Figure 10.2 in text), the bonds are nonpolar. CI$_4$ has tetrahedral geometry and is nonpolar. It has no dipole-dipole forces.

b) The electronegativities of C, H, and Cl are 2.5, 2.1, and 3.5, respectively. Consequently, CH$_3$Cl has three bonds that are nearly nonpolar (C–H) and one polar bond (C–Cl). The geometry of CH$_3$Cl is tetrahedral. Since the C–H bonds and the C–Cl bond are different, they do not cancel but sum to a net dipole moment. Therefore, the molecule is polar and has dipole-dipole forces.

c) Since the electronegativities of H and Cl are 2.1 and 3.5, respectively, HCl has a polar bond. The geometry of HCl is linear. The molecule is polar and has dipole-dipole forces.

SKILLBUILDER 12.5   Hydrogen Bonding

Which has the higher boiling point, HF or HCl? Why?

Solution:
HF has the higher boiling point. HF has hydrogen bonding as an intermolecular force, HCl does not.

SKILLBUILDER 12.6   Identifying Types of Crystalline Solids

Identify each of the following solids as molecular, ionic, or atomic.

a) NH$_3$(s)
b) CaO(s)
c) Kr(s)

Solution:
a) NH$_3$ is a molecular compound (nonmetal bonded to a nonmetal) and therefore forms a molecular solid.
b) CaO is an ionic solid (metal and nonmetal) and therefore forms an ionic solid.
c) Kr is a noble gas and therefore forms a nonbonding atomic solid.

SELF-TEST QUESTIONS

A. Match the following terms with the phrases below.

    boiling point   dipole-dipole force
    covalent atomic solid   dispersion force

Chapter 12

    dynamic equilibrium   metallic atomic solid
    evaporation   miscibility
    heat of fusion (ΔH_{fus})   nonbonding atomic solid
    heat of vaporization (ΔH_{vap})   nonvolatile
    hydrogen bond   normal boiling point
    instantaneous dipole   permanent dipole
    melting point   surface tension

1. Tendency of liquids to minimize their surface area
2. Physical change in which a substance is converted from its liquid form to its gaseous form
3. Characteristic of liquids that do not evaporate easily
4. Forward and reverse processes occur at the same rate
5. Temperature at which the vapor pressure of a liquid is equal to the pressure above it
6. Temperature at which the vapor pressure of a liquid is equal to 1 atmosphere
7. Amount of heat required to vaporize one mole of liquid
8. Temperature at which a solid turns into a liquid
9. Amount of heat required to melt one mole of a solid
10. London force
11. Fleeting charge separation
12. Force that exists between polar molecules
13. Type of dipole present in polar molecules
14. Ability of liquids to mix without separating
15. Attraction between a hydrogen atom and a fluorine, oxygen, or nitrogen atom of a neighboring molecule
16. Atomic solids held together by covalent bonds
17. Atomic solids held together by relatively weak dispersion forces
18. Atomic solids held together by metallic bonds

B. True/False

1. Liquids are easily compressed.
2. Solids have definite volume.
3. Water is more viscous than syrup.
4. Increasing the surface area of a liquid increases its rate of evaporation.
5. Vapor pressure increases with increasing temperature.
6. Evaporation is endothermic.
7. Freezing is exothermic.
8. Sublimation is a physical change in which a solid converts to a liquid and then to a gas.
9. Nonpolar molecules have dipole-dipole forces.
10. Water expands upon freezing.

Liquids, Solids, and Intermolecular Forces

C. Multiple Choice

1. Which of the following is characteristic of liquids?
   a) Liquids have low densities in comparison to gases
   b) Liquids are easily compressed
   c) Liquids have intermolecular forces intermediate in comparison to those of gases and solids
   d) Liquids have definite shapes

2. As intermolecular forces within a liquid increase,
   a) Viscosity decreases and surface tension increases
   b) Viscosity decreases and surface tension decreases
   c) Viscosity increases and surface tension increases
   d) Viscosity increases and surface tension decreases

3. Vapor pressure increases with
   a) Increasing temperature and increasing strength of intermolecular forces
   b) Increasing temperature and decreasing strength of intermolecular forces
   c) Decreasing temperature and increasing strength of intermolecular forces
   d) Decreasing temperature and decreasing strength of intermolecular forces

4. Sublimation is a physical change in which
   a) A solid converts to a liquid
   b) A solid converts to a gas
   c) A liquid converts to a gas
   d) A liquid converts to a solid

5. How much heat is required to vaporize 50.0 g of liquid acetone, C_3H_6O, at 82.3 °C? The heat of vaporization is 29.1 kJ/mol.
   a) 25.0 kJ
   b) 33.8 kJ
   c) 70.8 kJ
   d) 95.6 kJ

6. How many kilograms of $H_2O(s)$ at 0.00 °C will melt upon absorbing 764 kJ of heat? The heat of fusion is 6.02 kJ/mol.
   a) 0.127 kg
   b) 0.255 kg
   c) 0.334 kg
   d) 2.29 kg

7. Which of the following hydrocarbons has the strongest dispersion forces?
   a) CH_4
   b) C_2H_6
   c) C_3H_8
   d) C_4H_{10}

Chapter 12

8. Which of the following molecules has dipole-dipole forces?
    a) CH_3F
    b) CO_2
    c) CH_4
    d) CCl_4

9. Which of the following liquids is miscible with water?
    a) CBr_4
    b) CH_3CH_2OH
    c) C_6H_{14}
    d) C_9H_{20}

10. Which of the following is a molecular solid?
    a) NaCl
    b) Fe
    c) CaF_2
    d) I_2

D. Crossword Puzzle

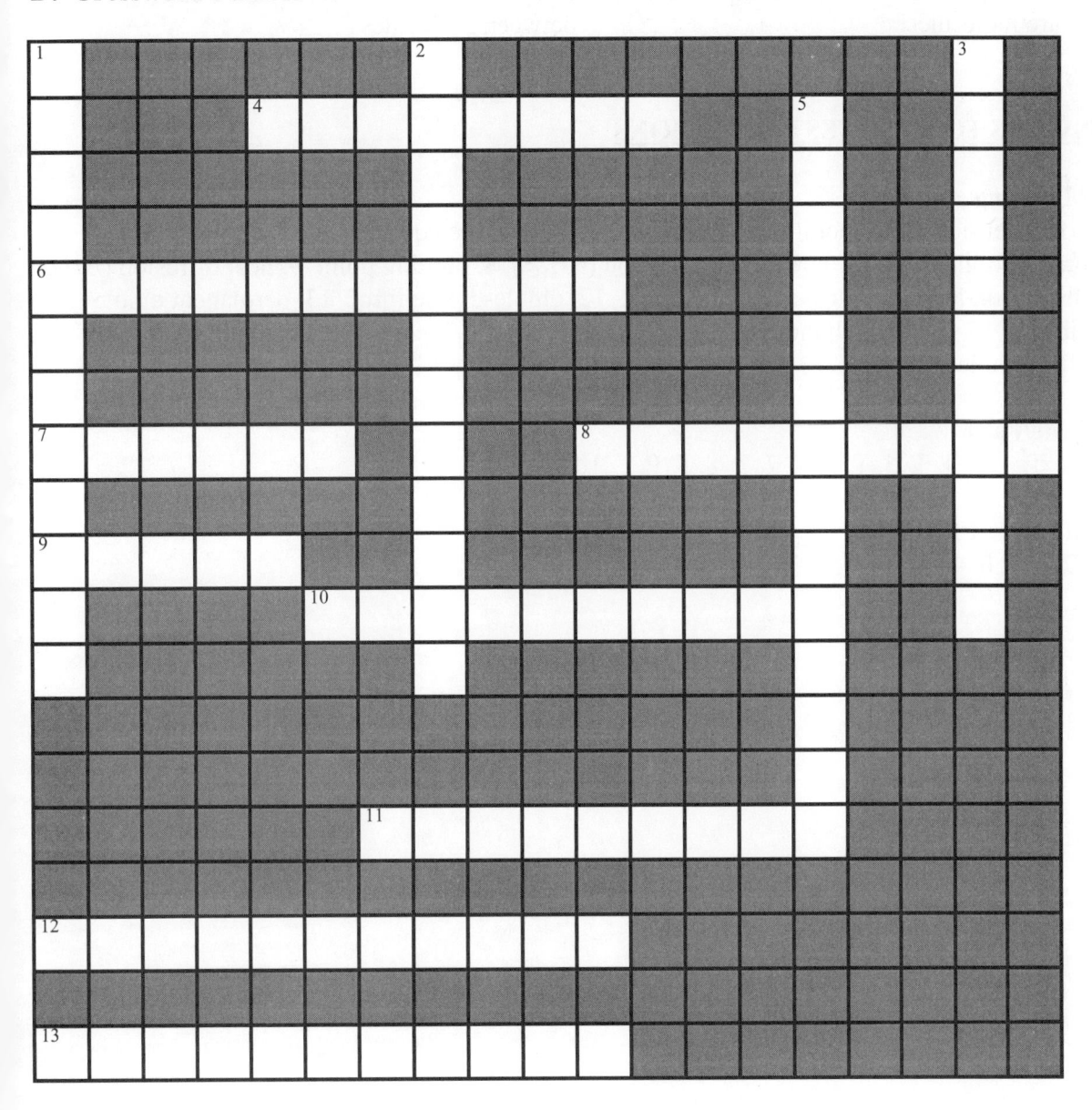

ACROSS

4. Characteristic of liquids that evaporate easily
6. Heat is absorbed
7. Type of solid whose composite units are individual atoms
8. Resistance of a liquid to flow
9. Type of solid whose composite units are formula units
10. Heat is released

DOWN

1. Physical change in which a substance is converted from its gaseous form to its liquid form
2. Physical change in which a substance is converted from a liquid to a gas
3. Physical change in which a substance is converted from its solid form directly into its gaseous form

Chapter 12

11. Type of solid whose composite units are molecules

5. Type of attractive force that exists between molecules

ANSWERS TO SELF-TEST QUESTIONS

A. Matching
1. surface tension 2. evaporation 3. nonvolatile 4. dynamic equilibrium 5. boiling point
6. normal boiling point 7. heat of vaporization (ΔH_{vap}) 8. melting point 9. heat of fusion (ΔH_{fus})
10. dispersion force 11. instantaneous dipole 12. dipole-dipole force 13. permanent dipole 14. miscibility 15. hydrogen bond 16. covalent atomic solid 17. nonbonding atomic solid 18. metallic atomic solid

B. True/False
1. F 2. T 3. F 4. T 5. T 6. T 7. T 8. F 9. F 10. T

C. Multiple Choice
1. c 2. c 3. b 4. b 5. a 6. d 7. d 8. a 9. b 10. d

D. Crossword Puzzle

	1						2							3	
	C						V							S	
	O				4V| O	L	A	T	I	L	E	5I		U	
	N						P					N		B	
	D						O					T		L	
6E	N	D	O	T	H	E	R	M	I	C		E		I	
	N						I					R		M	
	S						Z					M		A	
7A| T	O	M	I	C		A		8V| I	S	C	O	S	I	T	Y
	T						T					L		I	
9I	O	N	I	C			I					E		O	
	O				10E| X	O	T	H	E	R	M	I	C		N	
	N						N					U			
												L			
												A			
					11M	O	L	E	C	U	L	A	R		

183

Chapter 12

Solutions 13

CHAPTER OVERVIEW

Chapter 13 describes solutions and factors that affect solubility. Calculations pertaining to solution concentration expressed as mass percent and molarity are presented. Solution stoichiometry is examined. Colligative properties of solutions are discussed.

CHAPTER OBJECTIVES

After reading and studying the text, students should be able to:

1. List the two components of a solution.
2. Relate solubility to unsaturated, saturated, and supersaturated solutions.
3. Differentiate between a strong electrolyte solution and a nonelectrolyte solution.
4. Describe the effect of temperature on the solubility of a solid in a liquid.
5. Describe the effects of temperature and pressure on the solubility of a gas in a liquid.
6. Differentiate between a dilute solution and a concentrated solution.
7. Define mass percent as a solution concentration.
8. Calculate mass percent given the mass of solute and the mass of solvent.
9. Calculate mass percent given the mass of solute and volume and density of solution.
10. Use mass percent to convert between the mass of solute and the mass of solution.
11. Define molarity.
12. Calculate molarity given the mass or number of moles of solute and volume of solution.
13. Use molarity to convert between moles of solute and the volume of a solution.
14. Solve problems involving dilution of solutions.
15. Solve problems involving solution stoichiometry.
16. Describe the colligative properties of solutions.
17. Explain how the presence of a nonvolatile solute affects the freezing point and boiling point of the soluton.
18. Define molality.
19. Calculate molality given the mass or number of moles of solute and mass of solvent.
20. Calculate the freezing point depression from the molality of a solution and the freezing point depression constant.
21. Calculate the boiling point elevation from the molality of a solution and the boiling point elevation constant.
22. Describe the process of osmosis.

Chapter 13

CHAPTER IN REVIEW

- Solutions are homogeneous mixtures of two or more substances.

- Solutions have at least two components. The majority component is called the solvent. The minority component is called the solute.

- The solubility of a compound is the amount of the compound, usually in grams, that will dissolve in a specific amount of liquid.

- The solubility of a solid in a liquid tends to increase with increasing temperature.

- A saturated solution is one that holds the maximum amount of solute as described by the solubility of the compound. An unsaturated solution is one that holds less than the maximum amount of solute as described by the solubility of the compound. A supersaturated solution is one that holds more than the maximum amount of solute as described by the solubility of the compound.

- Strong electrolyte solutions contain solutes that dissociate into ions when dissolved.

- Nonelectrolyte solutions contain molecular solutes that do not form ions when dissolved.

- Recrystallization is a technique in which a solid is dissolved in a solvent at elevated temperature to form a saturated solution. The solution is slowly cooled to form crystals of higher purity.

- The solubility of a gas in a liquid increases with decreasing temperature and increasing partial pressure of the gas above the liquid.

- Dilute solutions contain a small amount of solute relative to the solvent. Concentrated solutions contain a large amount of solute relative to the solvent.

- Solution concentration expressed as mass percent is the number of grams of solute per 100 g solution.

- Mass percent can be used as a conversion factor between mass of solute and mass of solution.

- Molarity (M) is moles of solute per liter of solution.

- The molarity of a solution can be used as a conversion factor between the number of moles of solute and liters of solution.

- Stock solutions are concentrated forms of solutions. Less concentrated solutions are prepared by diluting stock solutions.

Solutions

- The freezing point of a solution containing a nonvolatile solute is lower than the freezing point of the pure solvent. The difference between the freezing point of the solvent and the freezing point of the solution is called freezing point depression.

- The boiling point of a solution containing a nonvolatile solute is higher than the boiling point of the pure solvent. The difference between the boiling point of the solution and the boiling point of the solvent is called boiling point elevation.

- Colligative properties of solutions are those that depend on the amount, not the type, of solute.

- Molality (*m*) is the number of moles of solute dissolved per kilogram of solvent.

- Osmosis is the flow of solvent from a lower-concentration solution to a higher-concentration solution.

- Semipermeable membranes are membranes that selectively allow some substances to pass through but not others.

- Osmotic pressure is the pressure required to stop osmotic flow.

SKILLBUILDER PROBLEMS AND SOLUTIONS

SKILLBUILDER 13.1   **Calculating Mass Percent**

Calculate the mass percent of a sucrose solution containing 11.3 g of sucrose and 412.1 mL of water. (Assume that the density of water is 1.00 g/mL.)

Given: 11.3 g sucrose
412.1 mL water

Find: mass percent

Equation and Conversion Factor:

$$\text{Mass percent} = \frac{\text{Mass solute}}{\text{Mass solution}} \times 100\%$$

This problem requires converting mL to g using the density of water as the conversion factor.

$$d(H_2O) = \frac{1.00 \text{ g}}{\text{mL}}$$

Chapter 13

Solution:
To find the mass percent, we simply substitute into the equation for mass percent. We need the mass of the solution, which is simply the mass of sucrose plus the mass of water. The mass of water is obtained from the volume of water by using density as a conversion factor.

$$\text{Mass H}_2\text{O} = 412.1 \text{ mL} \times \frac{1.00 \text{ g}}{\text{mL}} = 412.1 \text{ g}$$

$$\begin{aligned}\text{Mass solution} &= \text{Mass sucrose} + \text{Mass water} \\ &= 11.3 \text{ g} + 412.1 \text{ g} \\ &= 423.4 \text{ g}\end{aligned}$$

We then substitute the correct quantities into the equation.

$$\text{Mass percent} = \frac{\text{Mass solute}}{\text{Mass solution}} \times 100\%$$

$$= \frac{11.3 \text{ g}}{423.4 \text{ g}} \times 100\%$$

$$= 2.67\%$$

SKILLBUILDER 13.2   **Using Mass Percent in Calculations**

How much sucrose ($C_{12}H_{22}O_{11}$) in grams is contained in 355 mL (12 oz) of the soft drink in Example 13.2 in the text?

Given: 355 mL solution (soft drink)
11.5 % $C_{12}H_{22}O_{11}$ by mass

Find: g $C_{12}H_{22}O_{11}$

Conversion Factors:

$$\frac{11.5 \text{ g } C_{12}H_{22}O_{11}}{100 \text{ g solution}}$$

$$d(\text{solution}) = \frac{1.00 \text{ g}}{1 \text{ mL}}$$

Solutions

Solution Map:

$$mL\ solution \rightarrow g\ solution \rightarrow g\ C_{12}H_{22}O_{11}$$

$$\frac{1\ g\ solution}{1.00\ mL\ solution} \qquad \frac{11.5\ g\ C_{12}H_{22}O_{11}}{100\ g\ solution}$$

Solution:

$$355\ mL \times \frac{1.00\ g\ solution}{1\ mL\ solution} \times \frac{11.5\ g\ C_{12}H_{22}O_{11}}{100\ g\ solution} = 40.8\ g\ C_{12}H_{22}O_{11}$$

SKILLBUILDER 13.3   Calculating Molarity

Calculate the molarity of a solution made by putting 55.8 g of NaNO$_3$ into a beaker and diluting to 2.50 L.

Given: 55.8 g NaNO$_3$
2.50 L solution

Find: molarity (M)

Equation and Conversion Factor:

$$\text{Molarity (M)} = \frac{\text{Moles solute}}{\text{L solution}}$$

85.00 g NaNO$_3$ = 1 mol NaNO$_3$

Solution:
To calculate molarity, we simply substitute the correct values into the equation and compute the answer. However, we must first convert the amount of NaNO$_3$ from g to mol using the molar mass of NaNO$_3$.

$$55.8\ g\ NaNO_3 \times \frac{1\ mol\ NaNO_3}{85.00\ g\ NaNO_3} = 0.65\underline{6}47\ mol\ NaNO_3$$

$$\text{Molarity (M)} = \frac{\text{Moles solute}}{\text{L solution}}$$

$$= \frac{0.65\underline{6}47\ mol\ NaNO_3}{2.50\ L\ solution}$$

$$= 0.263\ M$$

Chapter 13

SKILLBUILDER 13.4 Using Molarity in Calculations

How much of a 0.225 M KCl solution contain 55.8 g KCl?

Given: 0.225 M KCl
55.8 g KCl

Find: L solution

Conversion Factors:

1 mol KCl = 74.55 g KCl

$$\frac{0.225 \text{ mol KCl}}{\text{L solution}}$$

Solution Map:
The main conversion factor is the molarity of the solution. We also need the molar mass of KCl.

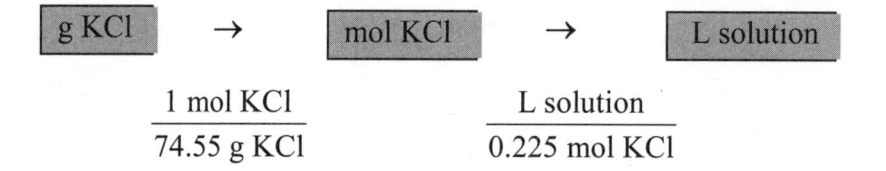

g KCl → mol KCl → L solution

$$\frac{1 \text{ mol KCl}}{74.55 \text{ g KCl}} \qquad \frac{\text{L solution}}{0.225 \text{ mol KCl}}$$

Solution:

$$55.8 \text{ g KCl} \times \frac{1 \text{ mol KCl}}{74.55 \text{ g KCl}} \times \frac{1 \text{ L solution}}{0.225 \text{ mol KCl}} = 3.33 \text{ L}$$

SKILLBUILDER 13.5 Ion Concentration

Determine the molar concentrations of Ca^{2+} and Cl^- in a 0.75 M $CaCl_2$ solution.

Given: 0.75 M $CaCl_2$

Find: molarity (M) of Ca^{2+} and Cl^-

Solution:

molarity of Ca^{2+} = 0.75 M

molarity of Cl^- = 2(0.75 M) = 1.5 M

Solutions

SKILLBUILDER 13.6 — Solution Dilution

How much 6.0 M NaNO$_3$ solution should be used to make 0.585 L of a 1.2 M NaNO$_3$ solution?

Given: $M_1 = 6.0$ M
$M_2 = 1.2$ M
$V_2 = 0.585$ L

Find: V_1

Equation:

$$M_1 V_1 = M_2 V_2$$

Solution:

$$M_1 V_1 = M_2 V_2$$

$$V_1 = \frac{M_2 V_2}{M_1}$$

$$= \frac{1.2 \text{ M} \times 0.585 \text{ L}}{6.0 \text{ M}}$$

$$= 0.12 \text{ L}$$

SKILLBUILDER 13.7 — Solution Stoichiometry

How many milliliters of 0.112 M Na$_2$CO$_3$ are necessary to completely react with 27.2 mL of 0.135 M HNO$_3$ according to the following reaction?

$$2 \text{ HNO}_3(aq) + \text{Na}_2\text{CO}_3(aq) \rightarrow \text{H}_2\text{O}(l) + \text{CO}_2(g) + 2 \text{ NaNO}_3(aq)$$

Given: 0.112 M Na$_2$CO$_3$
27.2 mL HNO$_3$ solution
0.135 M HNO$_3$

Find: mL Na$_2$CO$_3$ solution

Conversion Factors:

$$M(\text{Na}_2\text{CO}_3) = \frac{0.112 \text{ mol Na}_2\text{CO}_3}{\text{L Na}_2\text{CO}_3 \text{ solution}}$$

Chapter 13

$$M(HNO_3) = \frac{0.135 \text{ mol } HNO_3}{L \text{ } HNO_3 \text{ solution}}$$

2 mol HNO_3 ≡ 1 mol Na_2CO_3

1 mL = 10^{-3} L

Solution Map:

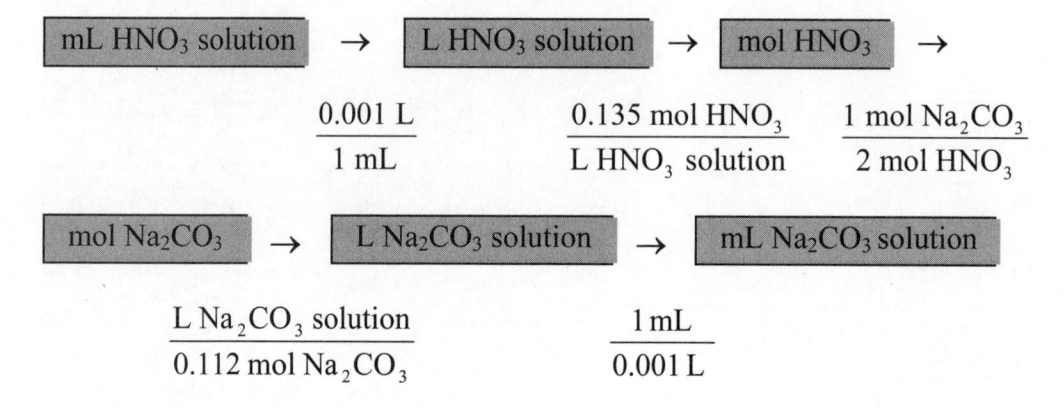

mL HNO_3 solution → L HNO_3 solution → mol HNO_3 →

$$\frac{0.001 \text{ L}}{1 \text{ mL}} \qquad \frac{0.135 \text{ mol } HNO_3}{L \text{ } HNO_3 \text{ solution}} \qquad \frac{1 \text{ mol } Na_2CO_3}{2 \text{ mol } HNO_3}$$

mol Na_2CO_3 → L Na_2CO_3 solution → mL Na_2CO_3 solution

$$\frac{L \text{ } Na_2CO_3 \text{ solution}}{0.112 \text{ mol } Na_2CO_3} \qquad \frac{1 \text{ mL}}{0.001 \text{ L}}$$

Solution:

$$27.2 \text{ mL } HNO_3 \text{ solution} \times \frac{0.001 \text{ L}}{1 \text{ mL}} \times \frac{0.135 \text{ mol } HNO_3}{L \text{ } HNO_3 \text{ solution}} \times \frac{1 \text{ mol } Na_2CO_3}{2 \text{ mol } HNO_3}$$

$$\times \frac{L \text{ } Na_2CO_3 \text{ solution}}{0.112 \text{ mol } Na_2CO_3} \times \frac{1 \text{ mL}}{0.001 \text{ L}} = 16.4 \text{ mL } Na_2CO_3 \text{ solution}$$

SKILLBUILDER PLUS

A 25.0-mL sample of HNO_3 solution requires 35.7 mL of 0.108 M Na_2CO_3 to completely react with all of the HNO_3 in the solution. What was the concentration of the HNO_3 solution?

$$2 \text{ } HNO_3(aq) + Na_2CO_3(aq) \rightarrow H_2O(l) + CO_2(g) + 2 \text{ } NaNO_3(aq)$$

Given: 25.0 mL HNO_3 solution
35.7 mL Na_2CO_3 solution
0.108 M Na_2CO_3

Find: M HNO_3

Solutions

Equation and Conversion Factors:

$$\text{Molarity (M)} = \frac{\text{Moles solute}}{\text{L solution}}$$

$$M(Na_2CO_3) = \frac{0.108 \text{ mol } Na_2CO_3}{L \; Na_2CO_3 \text{ solution}}$$

2 mol HNO_3 ≡ 1 mol Na_2CO_3

1 mL = 10^{-3} L

Solution Map:

| mL Na_2CO_3 solution | → | L Na_2CO_3 solution | → | mol Na_2CO_3 | → | mol HNO_3 |

$$\frac{0.001 \text{ L}}{1 \text{ mL}} \qquad \frac{0.108 \text{ mol } Na_2CO_3}{L \; Na_2CO_3 \text{ solution}} \qquad \frac{2 \text{ mol } HNO_3}{1 \text{ mol } Na_2CO_3}$$

Solution:
We follow the solution map to compute mol HNO_3.

$$35.7 \text{ mL } Na_2CO_3 \text{ solution} \times \frac{0.001 \text{ L}}{1 \text{ mL}} \times \frac{0.108 \text{ mol } Na_2CO_3}{L \; Na_2CO_3 \text{ solution}} \times \frac{2 \text{ mol } HNO_3}{1 \text{ mol } Na_2CO_3}$$

$$= 7.7\underline{1}12 \times 10^{-3} \text{ mol } HNO_3$$

Before we can substitute into the molarity equation, we must convert the volume of HNO_3 solution into the correct units.

$$25.0 \text{ mL} \times \frac{0.001 \text{ L}}{1 \text{ mL}} = 0.0250 \text{ L}$$

$$\text{Molarity (M)} = \frac{\text{Moles solute}}{\text{L solution}}$$

$$= \frac{7.7\underline{1}12 \times 10^{-3} \text{ mol } HNO_3}{0.0250 \text{ L solution}}$$

$$= 0.308 \text{ M}$$

Chapter 13

SKILLBUILDER 13.8   **Calculating Molality**

Calculate the molality (*m*) of a sucrose ($C_{12}H_{22}O_{11}$) solution containing 50.4 g sucrose and 0.332 kg of water.

Given: 50.4 g $C_{12}H_{22}O_{11}$
    0.332 kg H_2O

Find: molality (m)

Equation and Conversion Factor:

$$\text{Molality (m)} = \frac{\text{Moles solute}}{\text{Kilograms solvent}}$$

$$342.34 \text{ g } C_{12}H_{22}O_{11} = 1 \text{ mol } C_{12}H_{22}O_{11}$$

Solution:
To calculate molality, we simply substitute the correct values into the equation and compute the answer. However, we must first convert the amount of $C_{12}H_{22}O_{11}$ from g to mol using the molar mass of $C_{12}H_{22}O_{11}$.

$$50.4 \text{ g } C_{12}H_{22}O_{11} \times \frac{1 \text{ mol } C_{12}H_{22}O_{11}}{342.34 \text{ g } C_{12}H_{22}O_{11}} = 0.14722 \text{ mol } C_{12}H_{22}O_{11}$$

$$\text{Molality (m)} = \frac{\text{Moles solute}}{\text{Kilograms solvent}}$$

$$= \frac{0.14722 \text{ mol } C_{12}H_{22}O_{11}}{0.332 \text{ kg } H_2O}$$

$$= 0.443 \text{ m}$$

SKILLBUILDER 13.9   **Freezing Point Depression**

Calculate the freezing point of a 2.6 m sucrose solution. $K_f = 1.86 \dfrac{°C \text{ kg solvent}}{\text{mol solute}}$

Given: 2.6 m solution
    $K_f = 1.86 \dfrac{°C \text{ kg solvent}}{\text{mol solute}}$

Find: T_f

Solutions

Equations:

$$\Delta T_f = m \times K_f$$

$$\Delta T_f = T_{f\,solvent} - T_{f\,solution}$$

Solution:
To solve this problem, we substitute the values into the equation for freezing point depression and calculate ΔT_f.

$$\Delta T_f = m \times K_f$$

$$= 2.6 \, \frac{\text{mol solute}}{\text{kg solvent}} \times 1.86 \, \frac{°\text{C kg solvent}}{\text{mol solute}}$$

$$= 4.8 \, °\text{C}$$

Next we calculate the freezing point of the solution using the freezing point of pure water, 0.00 °C.

$$\Delta T_f = T_{f\,solvent} - T_{f\,solution}$$

$$T_{f\,solution} = T_{f\,solvent} - \Delta T_f$$

$$= 0.00 \, °\text{C} - 4.8 \, °\text{C}$$
$$= -4.8 \, °\text{C}$$

SKILLBUILDER 13.10 Boiling Point Elevation

Calculate the boiling point of a 3.5 m glucose solution. $K_b = 0.512 \, \frac{°\text{C kg solvent}}{\text{mol solute}}$

Given: 3.5 m solution

$$K_b = 0.512 \, \frac{°\text{C kg solvent}}{\text{mol solute}}$$

Find: T_b

Equations:

$$\Delta T_b = m \times K_f$$

$$\Delta T_b = T_{b\,solution} - T_{b\,solvent}$$

Solution:
To solve this problem, we substitute the values into the equation for boiling point elevation and calculate ΔT_b.

Chapter 13

$$\Delta T_b = m \times K_f$$

$$= 3.5 \, \frac{\text{mol solute}}{\text{kg solvent}} \times 0.512 \, \frac{°\text{C kg solvent}}{\text{mol solute}}$$

$$= 1.8 \, °\text{C}$$

Next we calculate the boiling point of the solution using the boiling point of pure water, 100.00 °C.

$$\Delta T_b = T_{b \, \text{solution}} - T_{b \, \text{solvent}}$$

$$T_{b \, \text{solution}} = \Delta T_b + T_{b \, \text{solvent}}$$

$$= 1.8 \, °\text{C} + 100.00 \, °\text{C} = 101.8 \, °\text{C}$$

SELF-TEST QUESTIONS

A. Match the following terms with the phrases below.

boiling point elevation
dilute solution
electrolyte solution
freezing point depression
mass percent

nonelectrolyte solution
osmotic pressure
saturated solution
stock solution
unsaturated solution

1. Solution holding the maximum amount of solute
2. Solution holding less than the maximum amount of solute
3. Solution containing a solute that dissociates into ions
4. Solution containing dissolved molecules
5. Solution containing a small amount of solute relative to the solvent
6. Number of grams of solute per 100 g solution
7. Concentrated form of a solution
8. Decrease in freezing point upon dissolving a solute in a solvent
9. Increase in boiling point upon dissolving a solute in a solvent
10. Pressure required to stop osmotic flow

B. True/False

1. Solutions are heterogeneous mixtures.
2. The concentration of an unsaturated solution is greater than its solubility.
3. Electrolyte solutions contain dissolved ions.
4. Gas solubility increases with increasing temperature.
5. Molarity is the number of moles solute per liter solvent.

6. Dilution is addition of solute to a solution.
7. Adding salt to water causes the boiling point to decrease.
8. Molality is the number of moles solute per liter solution.
9. Colligative properties depend on the amount of solute particles, not the type of solute particles.
10. Osmosis is the flow of solvent from a higher-concentration solution to a lower-concentration solution.

C. Multiple Choice

1. When sugar dissolves in water, water is termed the
   a) Solute
   b) Solvent
   c) Solution
   d) Soluble

2. A solution holding less than the maximum amount of solute is
   a) Unsaturated
   b) Saturated
   c) Supersaturated
   d) Dissolved

3. A solution is prepared by dissolving 18.1 g KCl in 90.0 water. Calculate the mass percent of the solution.
   a) 16.7%
   b) 20.1%
   c) 49.7%
   d) 83.3%

4. How many grams of $NaNO_3$ are in 55.0 g of a solution that has a concentration of 8.50% by mass?
   a) 4.68 g
   b) 6.47 g
   c) 15.5 g
   d) 46.5 g

5. What is the molarity of a solution prepared by dissolving 2.26 g Na_2CO_3 in water and diluting to 500.0 mL?
   a) 0.0426 M
   b) 0.0479 M
   c) 0.0452 M
   d) 0.0545 M

Chapter 13

6. How many grams of K_2SO_4 are in 75.0 mL of a 0.133 M solution?
   a) 0.309 g
   b) 0.564 g
   c) 1.74 g
   d) 9.98 g

7. A 25.0-mL sample of a 0.502 M solution is diluted to 200.0 mL. What is the molarity of the dilute solution?
   a) 0.0126 M
   b) 0.0201 M
   c) 0.0402 M
   d) 0.0628 M

8. What volume of 0.102 M NaOH will completely neutralize 50.0 mL of 0.108 M H_3PO_4?

   $$3\ NaOH(aq)\ +\ H_3PO_4(aq) \rightarrow\ Na_3PO_4(aq)\ +\ 3\ H_2O(l)$$

   a) 47.2 mL
   b) 52.9 mL
   c) 159 mL
   d) 142 mL

9. What is the molality of a solution prepared by dissolving 15.0 g ethanol (C_2H_5OH) in 75.0 g H_2O?
   a) 0.109 m
   b) 0.200 m
   c) 0.325 m
   d) 4.34 m

10. What is the freezing point of a solution prepared by dissolving 2.60 g dextrose, $C_6H_{12}O_6$, in 42.0 g H_2O? The freezing point of water is 0.00 °C.

    K_f for water = 1.86 $\dfrac{\text{°C kg solvent}}{\text{mol solute}}$

    a) 0.34 °C
    b) 0.64 °C
    c) −0.34 °C
    d) −0.64 °C

Solutions

D. Crossword Puzzle

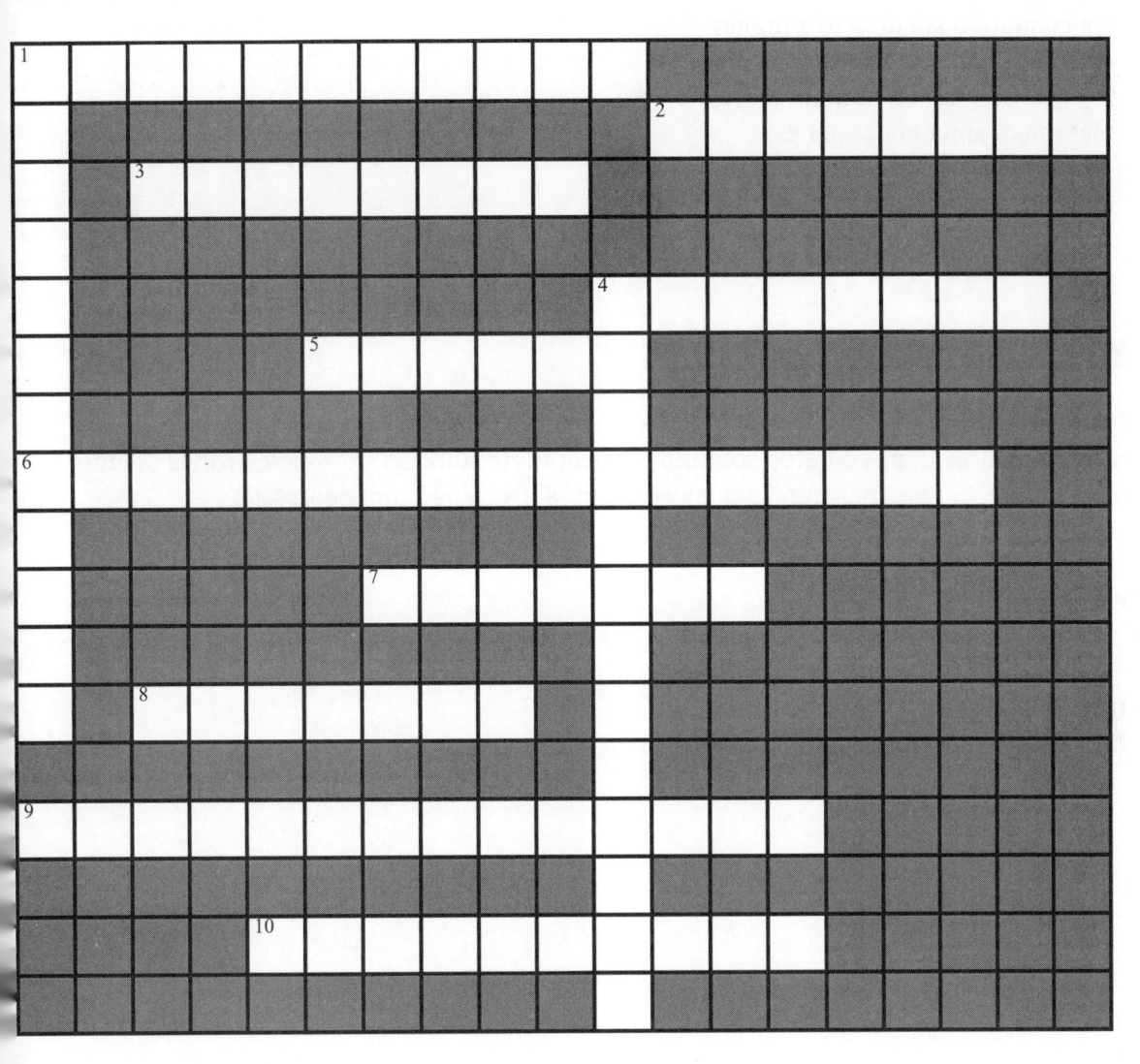

ACROSS

1. Type of property that depends on the amount of solute and not the type of solute
2. Number of moles of solute per kilogram of solvent
3. Number of moles of solute per liter of solution
4. Homogeneous mixture of two or more substances
5. Minority component of a solution
6. Purification technique based on the temperature dependence of solubility
7. Majority component of a solution

DOWN

1. Type of solution containing a large amount of solute relative to the solvent
4. Type of membrane that selectively allows some substances to pass through but not others

Chapter 13

8. Flow of a solvent from a lower-concentration solution to a higher-concentration solution
9. Type of solution holding more than the maximum amount of solute
10. Amount of compound that dissolves in a certain amount of liquid

ANSWERS TO SELF-TEST QUESTIONS

A. Matching
1. saturated solution 2. unsaturated solution 3. electrolyte solution 4. nonelectrolyte solution
5. dilute solution 6. mass percent 7. stock solution 8. freezing point depression
9. boiling point elevation 10. osmotic pressure

B. True/False
1. F 2. F 3. T 4. F 5. F 6. F 7. F 8. F 9. T 10. F

C. Multiple Choice
1. b 2. a 3. a 4. a 5. a 6. c 7. d 8. c 9. d 10. d

D. Crossword Puzzle

¹C	O	L	L	I	G	A	T	I	V	E				
O										²M	O	L	A	L	I	T	Y
N		³M	O	L	A	R	I	T	Y					
C														
E							⁴S	O	L	U	T	I	O	N
N				⁵S	O	L	U	T	E					
T							M							
⁶R	E	C	R	Y	S	T	A	L	L	I	Z	A	T	I	O	N
A							P							
T				⁷S	O	L	V	E	N	T				
E							R							
D		⁸O	S	M	O	S	I	S						
							M							
							E							
⁹S	U	P	E	R	S	A	T	U	R	A	T	E	D	
							B							
			¹⁰S	O	L	U	B	I	L	I	T	Y		
							E							

Chapter 13

Acids and Bases   14

CHAPTER OVERVIEW

The topic of Chapter 14 is acids and bases. Arrhenius and Brønsted-Lowry definitions are discussed. Reactions of acids and bases are described. The concept of pH is examined.

CHAPTER OBJECTIVES

After reading and studying the text, students should be able to:

1. List the properties of acids and bases.
2. Give examples of acids and bases.
3. State the Arrhenius definition of acids and bases.
4. State the Brønsted-Lowry definition of acids and bases.
5. Identify Brønsted-Lowry acids and bases and their conjugates.
6. Write balanced equations for neutralization reactions.
7. Write balanced equations for acids reacting with metals.
8. Write balanced equations for acids reacting with metal oxides.
9. Describe the titration process.
10. Solve problems involving acid-base titrations.
11. Differentiate between strong acids and weak acids.
12. Give examples of strong and weak acids.
13. Differentiate between strong and weak bases.
14. Give examples of strong and weak bases.
15. Calculate the hydronium ion concentration, given the molarity of a solution of a strong monoprotic acid.
16. Calculate the hydroxide ion concentration given, the molarity of a solution of a strong base.
17. Write the expression for the ion product constant for water.
18. State the relative concentrations of hydronium ion and hydroxide ion in acidic, neutral, and basic solutions.
19. Use K_w and the concentration of hydronium ion to calculate hydroxide ion concentration.
20. Use K_w and the concentration of hydroxide ion to calculate hydronium ion concentration.
21. Calculate pH from hydronium ion concentration.
22. Calculate hydronium ion concentration from pH.
23. State the components of a buffer.
24. Explain how buffers resist changes in pH.
25. Write equations for the chemical reactions by which acid rain forms in the atmosphere.
26. Describe the effects of acid rain and how the problem of acid rain is being addressed in the United States.

CHAPTER IN REVIEW

- Acids have a sour taste, dissolve many metals, and turn litmus paper red.

- Bases have a bitter taste, a slippery feel, and turn litmus paper blue.

- The Arrhenius definition of an acid is a substance that produces H^+ in aqueous solution.

- The Arrhenius definition of a base is a substance that produces OH^- in aqueous solution.

- In water, H^+ ions associate with water to form hydronium ions, H_3O^+.

- The Brønsted-Lowry definition of an acid is a proton (H^+) donor. The Brønsted-Lowry definition of a base is a proton (H^+) acceptor.

- In the Brønsted-Lowry definition, acids and bases always occur together. A proton (H^+) is transferred from the acid to the base.

- Amphoteric substances are substances that can act as acids or bases. Water is amphoteric.

- Conjugate acid-base pairs are substances that are related to each other by the transfer of a proton (H^+).

- Neutralization reactions are acid-base reactions. The products are generally a salt and water. Reaction of acids with carbonate or bicarbonate salts produce water, carbon dioxide, and a salt.

- Reactions between an acid and a metal usually produce hydrogen gas and a dissolved salt containing the cation of the metal.

- Acids react with metal oxides to produce water and a dissolved salt containing the cation of the metal.

- Titration is a laboratory technique in which a reactant in a solution of known concentration is reacted with another reactant in a solution of unknown concentration.

- The endpoint of a titration is the experimentally determined point at which the reactants are in stoichiometric proportions. Endpoints are often detected with visual indicators, substances that change color.

- Strong acids completely ionize in solution and are strong electrolytes. Weak acids do not completely ionize in solution and are weak electrolytes.

Acids and Bases

- The degree to which an acid is strong or weak depends on the attraction between the anion of the acid (its conjugate base) and hydrogen ion.

- Monoprotic acids contain one ionizable hydrogen. Diprotic acids contain two ionizable hydrogens.

- Strong bases completely dissociate in aqueous solution. Most weak bases react with water, causing water molecules to ionize.

- The ion product constant for water, K_w, is the product of the hydronium ion concentration and the hydroxide ion concentration.

- In neutral solutions, $[H_3O^+] = [OH^-] = 1.0 \times 10^{-7}$ M.

- In acidic solutions, $[H_3O^+] > [OH^-]$.

- In basic solutions, $[H_3O^+] < [OH^-]$.

- In all aqueous solutions at 25 °C, $K_w = [H_3O^+][OH^-] = 1.0 \times 10^{-14}$.

- The pH scale is a logarithmic scale used to specify the acidity of a solution.

- In neutral solutions, pH = 7. In acidic solutions, pH < 7. In basic solutions, pH > 7.

- Buffers are solutions that contain significant amounts of both a weak acid and its conjugate base or a weak base and its conjugate acid.

- Buffers resist changes in pH. The acid component of the buffer neutralizes added base. The base component of the buffer neutralizes added acid.

- Acid rain is the result of sulfur oxides and nitrogen oxides emitted by fossil fuel combustion. These oxides react with water to form sulfuric acid and nitric acid, which then fall as acid rain.

SKILLBUILDER PROBLEMS AND SOLUTIONS

SKILLBUILDER 14.1   Identifying Brønsted-Lowry Acids and Bases and Their Conjugates

In each of the following reactions, identify the Brønsted-Lowry acid, the Brønsted-Lowry base, the conjugate acid, and the conjugate base.

a) $C_5H_5N(aq) + H_2O(l) \rightleftharpoons C_5H_5NH^+(aq) + OH^-(aq)$

Chapter 14

b) $HNO_3(aq) + H_2O(l) \rightarrow H_3O^+(aq) + NO_3^-(aq)$

Solution:
a) Since H_2O donates a proton to C_5H_5N in this reaction, it is the acid (proton donor). After H_2O donates the proton, it becomes OH^-, the conjugate base. Since C_5H_5N accepts the proton, it is the base (proton acceptor). After C_5H_5N accepts the proton it becomes $C_5H_5NH^+$, the conjugate acid.

$$C_5H_5N(aq) + H_2O(l) \rightleftharpoons C_5H_5NH^+(aq) + OH^-(aq)$$
$$\text{Base} \qquad \text{Acid} \qquad \text{Conjugate acid} \qquad \text{Conjugate base}$$

b) Since HNO_3 donates a proton to H_2O in this reaction, it is the acid (proton donor). After HNO_3 donates the proton, it becomes NO_3^-, the conjugate base. Since H_2O accepts the proton, it is the base (proton acceptor). After H_2O accepts the proton it becomes H_3O^+, the conjugate acid.

$$HNO_3(aq) + H_2O(l) \rightarrow H_3O^+(aq) + NO_3^-(aq)$$
$$\text{Acid} \qquad \text{Base} \qquad \text{Conjugate acid} \qquad \text{Conjugate base}$$

SKILLBUILDER 14.2   **Writing Equations for Neutralization Reactions**

Write a molecular equation for the reaction that occurs between aqueous H_3PO_4 and aqueous NaOH. Hint: H_3PO_4 is a triprotic acid, meaning that 1 mol of H_3PO_4 requires 3 mol of OH^- to completely react with it.

Solution:
We must identify the acid and the base and know that they react to form water and a salt. One mole H_3PO_4 requires three moles NaOH. We first write the skeletal equation.

$$H_3PO_4(aq) + NaOH(aq) \rightarrow H_2O(l) + Na_3PO_4(aq)$$

We then balance the equation.

$$H_3PO_4(aq) + 3\ NaOH(aq) \rightarrow 3\ H_2O(l) + Na_3PO_4(aq)$$

SKILLBUILDER 14.3   **Writing Equations for Acid Reactions**

Write an equation for each of the following.

a) The reaction of hydrochloric acid with strontium metal
b) The reaction of hydroiodic acid with barium oxide

Acids and Bases

Solution:

a) The reaction of hydrochloric acid with strontium metal forms hydrogen and a salt. The salt contains the ionized form of the metal (Sr^{2+}) as the cation and the anion of the acid (Cl^-) as the anion. We first write the skeletal equation.

$$HCl(aq) + Sr(s) \rightarrow H_2(g) + SrCl_2(aq)$$

We then balance the equation.

$$2\ HCl(aq) + Sr(s) \rightarrow H_2(g) + SrCl_2(aq)$$

b) The reaction of hydroiodic acid and barium hydroxide forms water and a salt. The salt contains the cation from the metal oxide (Ba^{2+}) and the anion of the acid (I^-). We first write the skeletal equation.

$$HI(aq) + BaO(s) \rightarrow H_2O(l) + BaI_2(aq)$$

We then balance the equation.

$$2\ HI(aq) + BaO(s) \rightarrow H_2O(l) + BaI_2(aq)$$

SKILLBUILDER 14.4   **Acid-Base Titration**

The titration of a 20.0-mL sample of an H_2SO_4 solution of unknown concentration requires 22.87 mL of a 0.158 M KOH solution to reach the endpoint. What is the concentration of the unknown H_2SO_4 solution?

Given: 20.0 mL H_2SO_4 solution
22.87 mL KOH solution
0.158 M KOH

Find: M H_2SO_4

Equation and Conversion Factors:

$$\text{Molarity (M)} = \frac{\text{Moles solute}}{\text{L solution}}$$

$$\text{M (KOH)} = \frac{0.158\ \text{mol KOH}}{\text{L KOH solution}}$$

2 mol KOH ≡ 1 mol H_2SO_4

1 mL = 10^{-3} L

Chapter 14

Balanced Equation:

$$H_2SO_4(aq) + 2\ KOH(aq) \rightarrow 2\ H_2O(l) + K_2SO_4(aq)$$

Solution Map:

mL KOH solution → L KOH solution → mol KOH → mol H₂SO₄

$$\frac{0.001\ L}{1\ mL} \qquad \frac{0.158\ mol\ KOH}{L\ KOH\ solution} \qquad \frac{1\ mol\ H_2SO_4}{2\ mol\ KOH}$$

Solution:
We follow the solution map to compute mol H₂SO₄.

$$22.87\ mL\ KOH\ solution \times \frac{0.001\ L}{1\ mL} \times \frac{0.158\ mol\ KOH}{L\ KOH\ solution} \times \frac{1\ mol\ H_2SO_4}{2\ mol\ KOH}$$

$$= 1.8067 \times 10^{-3}\ mol\ HNO_3$$

Before we can substitute into the molarity equation, we must convert the volume of H₂SO₄ solution into the correct units.

$$20.0\ mL \times \frac{0.001\ L}{1\ mL} = 0.0200\ L$$

$$\text{Molarity (M)} = \frac{\text{Moles solute}}{\text{L solution}}$$

$$= \frac{0.0018067\ mol\ H_2SO_4}{0.0200\ L\ solution}$$

$$= 0.0903\ M$$

The unknown H₂SO₄ solution has a concentration of 0.0903 M.

SKILLBUILDER 14.5 Determining [H₃O⁺] in Acid Solutions

What is the H₃O⁺ concentration in each of the following solutions?

a) 0.50 M HCHO₂
b) 1.25 M HI
c) 0.75 M HF

Acids and Bases

Solution:

a) Since HCHO₂ is a weak acid, it partially ionizes. The concentration of H₃O⁺ will be less than 0.50 M.

b) Since HI is a strong acid, it completely ionizes. The concentration of H₃O⁺ will be 1.25 M.

d) Since HF is a weak acid, it partially ionizes. The concentration of H₃O⁺ will be less than 0.75 M.

SKILLBUILDER 14.6   Determining [OH⁻] in Base Solutions

What is the OH⁻ concentration in each of the following solutions?

a) 0.055 M Ba(OH)₂
b) 1.05 M C₅H₅N
c) 0.45 M NaOH

Solution:

a) Since Ba(OH)₂ is a strong base, it completely dissociates into Ba²⁺ and 2OH⁻ in solution. Ba(OH)₂ forms 2 mol of OH⁻ for every 1 mol of Ba(OH)₂. Consequently, the concentration of OH⁻ will be twice the concentration of Ba(OH)₂.

$$[OH^-] = 2(0.055 \text{ M}) = 0.11 \text{ M}$$

b) Since C₅H₅N is a weak base, it partially ionizes water. The concentration of OH⁻ will be less than 1.05 M.

c) Since NaOH is a strong base, it completely dissociates into Na⁺ and OH⁻ in solution. The concentration of OH⁻ will be 0.45 M.

SKILLBUILDER 14.7   Using K_w in Calculations

Calculate [H₃O⁺] in each of the following solutions and determine whether the solution is acidic, basic, or neutral.

a) [OH⁻] = 1.5 × 10⁻² M
b) [OH⁻] = 1.0 × 10⁻⁷ M
c) [OH⁻] = 8.2 × 10⁻¹⁰ M

Solution:

a) To find [H₃O⁺] we use the ion product constant.

$$[H_3O^+][OH^-] = K_w = 1.0 \times 10^{-14}$$

Chapter 14

We substitute the given value for [OH⁻] and solve the equation for [H₃O⁺].

$$[H_3O^+][1.5 \times 10^{-2}] = 1.0 \times 10^{-14}$$

$$[H_3O^+] = \frac{1.0 \times 10^{-14}}{1.5 \times 10^{-2}} = 6.7 \times 10^{-13} \text{ M}$$

Since [H₃O⁺] < 1.0 × 10⁻⁷ M and [OH⁻] > 1.0 × 10⁻⁷ M, the solution is basic.

b) We again substitute the given value for [OH⁻] and solve the ion product equation for [H₃O⁺].

$$[H_3O^+][1.0 \times 10^{-7}] = 1.0 \times 10^{-14}$$

$$[H_3O^+] = \frac{1.0 \times 10^{-14}}{1.0 \times 10^{-7}} = 1.0 \times 10^{-7} \text{ M}$$

Since [H₃O⁺] = 1.0 × 10⁻⁷ M and [OH⁻] = 1.0 × 10⁻⁷ M, the solution is neutral.

c) We again substitute the given value for [OH⁻] and solve the ion product equation for [H₃O⁺].

$$[H_3O^+][8.2 \times 10^{-10}] = 1.0 \times 10^{-14}$$

$$[H_3O^+] = \frac{1.0 \times 10^{-14}}{8.2 \times 10^{-10}} = 1.2 \times 10^{-5} \text{ M}$$

Since [H₃O⁺] > 1.0 × 10⁻⁷ M and [OH⁻] < 1.0 × 10⁻⁷ M, the solution is acidic.

SKILLBUILDER 14.8 Calculating pH from [H₃O⁺]

Calculate the pH of each of the following solutions and indicate whether the solution is acidic or basic.

a) [H₃O⁺] = 9.5 × 10⁻⁹ M
b) [H₃O⁺] = 6.1 × 10⁻³ M

Solution:
To calculate pH, simply substitute the given [H₃O⁺] into the pH equation.

$$\text{pH} = -\log [H_3O^+] \times 10^{-9}]$$
a) = −(−8.02)
   = −log [9.5
   = 8.02

210

Since the pH > 7, this solution is basic.

b) pH = $-\log [H_3O^+]$
   = $-\log [6.1 \times 10^{-3}]$
   = $-(-2.21)$
   = 2.21

Since the pH < 7, this solution is acidic.

SKILLBUILDER PLUS

Calculate the pH of a solution with $[OH^-] = 1.3 \times 10^{-2}$ M and indicate whether the solution is acidic or basic. Hint: Begin by using K_w to find $[H_3O^+]$.

Solution:
We find $[H_3O^+]$ using the ion product constant.

$$[H_3O^+][OH^-] = K_w = 1.0 \times 10^{-14}$$

We substitute the given value for $[OH^-]$ and solve the equation for $[H_3O^+]$.

$$[H_3O^+][1.3 \times 10^{-2}] = 1.0 \times 10^{-14}$$

$$[H_3O^+] = \frac{1.0 \times 10^{-14}}{1.3 \times 10^{-2}} = 7.7 \times 10^{-13}$$

To calculate pH, we substitute the value of $[H_3O^+]$ into the pH equation.

pH = $-\log [H_3O^+]$
   = $-\log [7.7 \times 10^{-13}]$
   = $-(-12.11)$
   = 12.11

Since the pH > 7, this solution is basic.

SKILLBUILDER 14.9   Calculating [H₃O⁺] from pH

Calculate the H_3O^+ concentration for a solution with a pH of 8.37.

Solution:
To find the $[H_3O^+]$ from pH, we must undo the log function. Use either Method 1 or Method 2 shown below.

Chapter 14

Method 1: Inverse Log Function   **Method 2: 10^x function**

$$pH = -\log[H_3O^+]$$
$$8.37 = -\log[H_3O^+]$$
$$-8.37 = \log[H_3O^+]$$
$$\text{invlog}(-8.37) = \text{invlog}(\log[H_3O^+])$$
$$\text{invlog}(-8.37) = [H_3O^+]$$
$$[H_3O^+] = 4.3 \times 10^{-9} \text{ M}$$

$$pH = -\log[H_3O^+]$$
$$8.37 = -\log[H_3O^+]$$
$$-8.37 = \log[H_3O^+]$$
$$10^{-8.37} = 10^{\log[H_3O^+]}$$
$$10^{-8.37} = [H_3O^+]$$
$$[H_3O^+] = 4.3 \times 10^{-9} \text{ M}$$

SKILLBUILDER PLUS

Calculate the OH$^-$ concentration for a solution with a pH of 3.66.

Solution:
First we must find $[H_3O^+]$ from the pH. Use either Method 1 or Method 2 shown below.

Method 1: Inverse Log Function   **Method 2: 10^x function**

$$pH = -\log[H_3O^+]$$
$$3.66 = -\log[H_3O^+]$$
$$-3.66 = \log[H_3O^+]$$
$$\text{invlog}(-3.66) = \text{invlog}(\log[H_3O^+])$$
$$\text{invlog}(-3.66) = [H_3O^+]$$
$$[H_3O^+] = 2.187 \times 10^{-4} \text{ M}$$

$$pH = -\log[H_3O^+]$$
$$3.66 = -\log[H_3O^+]$$
$$-3.66 = \log[H_3O^+]$$
$$10^{-3.66} = 10^{\log[H_3O^+]}$$
$$10^{-3.66} = [H_3O^+]$$
$$[H_3O^+] = 2.187 \times 10^{-4} \text{ M}$$

To find [OH$^-$] we use the ion product constant.

$$[H_3O^+][OH^-] = K_w = 1.0 \times 10^{-14}$$

We substitute the value for $[H_3O^+]$ and solve the equation for $[OH^-]$.

$$[2.187 \times 10^{-4}][OH^-] = 1.0 \times 10^{-14}$$

$$[OH^-] = \frac{1.0 \times 10^{-14}}{2.187 \times 10^{-4}} = 4.6 \times 10^{-11} \text{ M}$$

SELF-TEST QUESTIONS

A. Match the following terms with the phrases below.

    acid rain   Arrhenius base
    acidic solution   basic solution

Arrhenius acid
Brønsted-Lowry base
conjugate acid-base pair
equivalence point
ion product constant (K_w)
logarithmic scale
monoprotic acid
neutral solution

Brønsted-Lowry acid
strong acid
strong base
strong electrolyte
titration
weak acid
weak base
weak electrolyte

1. Substance that produces H^+ ions in aqueous solution
2. Substance that produces OH^- ions in aqueous solution
3. Two substances related to each other by the transfer of a proton
4. Proton (H^+ ion) donor
5. Proton (H^+ ion) acceptor
6. Laboratory procedure in which a solution of known concentration is reacted with another reactant in a solution of unknown concentration
7. Point in a titration at which the reactants are in stoichiometric proportions
8. Acid that completely ionizes in solution
9. Substance whose aqueous solutions are good conductors of electricity
10. Acid that contains only one ionizable proton
11. Acid that does not completely ionize in solution
12. Substance whose aqueous solutions are poor conductors of electricity
13. Base that completely dissociates in solution
14. Base that does not completely dissociate in solution
15. Product of the H_3O^+ ion concentration and the OH^- ion concentration in aqueous solution
16. Solution in which the H_3O^+ ion concentration and the OH^- ion concentration are equal
17. Solution in which the H_3O^+ ion concentration is greater than the OH^- ion concentration
18. Solution in which the OH^- ion concentration is greater than the H_3O^+ ion concentration
19. Scale based on powers of ten
20. Rain with pH values lower than normal

B. True/False

1. Acids turn litmus paper red.
2. Bases have a slippery feel.
3. Arrhenius bases produce hydronium ions in water.
4. The conjugate acid of ammonia is ammonium ion.
5. Reactions between acids and metals usually produce hydrogen gas.
6. Hydrochloric acid is a strong acid.
7. Ammonia is a strong base.
8. Water is amphoteric.
9. A solution becomes more acidic as its pH increases.
10. Buffers contain a strong acid and a strong base.

Chapter 14

C. **Multiple Choice**

1. A Brønsted-Lowry base is defined as a
   a) Substance that produces H_3O^+ in aqueous solution
   b) Substance that produces OH^- in aqueous solution
   c) Proton (H^+ ion) donor
   d) Proton (H^+ ion) acceptor

2. What is the conjugate base of $H_2PO_3^-$?
   a) H_3PO_3
   b) HPO_3^-
   c) HPO_3^{2-}
   d) PO_3^{3-}

3. Which of the following is the correct formula for the salt produced by the complete neutralization of phosphoric acid, H_3PO_4, by potassium hydroxide?
   a) KPO_4
   b) K_2PO_4
   c) K_3PO_4
   d) $KOPO_4$

4. What are the products of the reaction of hydrochloric acid with barium oxide?
   a) $BaCl$ and H_2O
   b) $BaCl_2$ and H_2O
   c) $BaOCl_2$ and H_2
   d) $BaOCl$ and H_2

5. Which of the following are the products of the reaction of hydroiodic acid with calcium metal?
   a) CaI and H_2O
   b) CaI and H_2
   c) CaI_2 and H_2O
   d) CaI_2 and H_2

6. What are the products of the reaction of hydrochloric acid with potassium hydrogen carbonate?
   a) KCl, H_2O, and CO_3
   b) KCl, H_2O, and CO_2
   c) KCl, H_2, and CO_2
   d) K_2CO_3 and H_2O

Acids and Bases

7. Oxalic acid, $H_2C_2O_4$, is diprotic. The titration of 25.00 mL of an $H_2C_2O_4$ solution required 32.42 mL of a 0.1032 M NaOH solution to reach the endpoint. What is the molarity of the $H_2C_2O_4$ solution?

$$H_2C_2O_4(aq) + 2\,NaOH(aq) \rightarrow Na_2C_2O_4(aq) + 2\,H_2O(l)$$

a) 0.06691 M
b) 0.07958 M
c) 0.1338 M
d) 0.1592 M

8. How do the molar concentrations of H_3O^+ and OH^- compare in a neutral aqueous solution?
   a) $[H_3O^+] > [OH^-]$
   b) $[H_3O^+] < [OH^-]$
   c) $[H_3O^+] = [OH^-]$
   d) There are no H_3O^+ or OH^- ions present in a neutral solution

9. What is the pH of a 0.15 M HCl solution?
   a) −1.41
   b) −0.82
   c) 0.82
   d) 1.41

10. What is the molar concentration of H_3O^+ in a solution with a pH of 12.00?
    a) 1.0×10^{-12}
    b) 1.0×10^{-2}
    c) 1.0×10^{2}
    d) 1.0×10^{12}

Chapter 14

D. Crossword Puzzle

ACROSS

4. Formation of ions
6. Substance that resists pH changes by neutralizing added acid or added base
7. − log [H$_3$O$^+$]
8. Reaction of an acid with a base
11. Substance that changes color with acidity level
12. Characteristic of substances that can act as acids or bases
13. Type of acid that contains a −COOH group

DOWN

1. Type of acid that contains two ionizable protons
2. Ion formed when H$^+$ bonds to water
3. Swedish chemist who proposed molecular definitions of acids and bases
5. Ionic compound
9. Organic bases found in plants
10. Breaking apart of a substance into its constituent ions
14. Substance that produces hydroxide ions in aqueous solution

Acids and Bases

ANSWERS TO SELF-TEST QUESTIONS

A. Matching
1. Arrhenius acid 2. Arrhenius base 3. conjugate acid-base pair 4. Brønsted-Lowry acid
5. Brønsted-Lowry base 6. titration 7. equivalence point 8. strong acid 9. strong electrolyte
10. monoprotic acid 11. weak acid 12. weak electrolyte 13. strong base 14. weak base 15. ion product constant (K_w) 16. neutral solution 17. acidic solution 18. basic solution
19. logarithmic scale 20. acid rain

B. True/False
1. T 2. T 3. F 4. T 5. T 6. T 7. F 8. T 9. F 10. F

C. Multiple Choice
1. d 2. c 3. c 4. b 5. d 6. b 7. a 8. c 9. c 10. a

D. Crossword Puzzle

				¹D								²H			³A
		⁴I	O	N	I	Z	A	T	I	O	N	Y			R
				P					⁵S			D			R
⁶B	U	F	F	E	R				A			R		⁷p	H
				O					L			O			E
		⁸N	E	U	T	R	A	L	I	Z	⁹A	T	I	O	N
	¹⁰D			I					L		T				I
	¹¹I	N	D	I	C	A	T	O	R		L		U		U
	S								K		A		M		S
	S								A						
	O								L						
	C		¹²A	M	P	H	O	T	E	R	I	C			
	I								D						
¹³C	A	R	¹⁴B	O	X	Y	L	I	C						
	T		A						S						
	I		S												
	O		E												
	N														

Chapter 14

Chemical Equilibrium 15

CHAPTER OVERVIEW

Chapter 15 looks at chemical equilibrium. Reaction rates and dynamic equilibrium are discussed. Equilibrium constant expressions and equilibrium calculations are presented. Le Châtelier's principle is examined. Reaction pathways and catalysts are discussed.

CHAPTER OBJECTIVES

After reading and studying the text, students should be able to:

1. Describe the process of dynamic equilibrium.
2. Define reaction rate.
3. Use collision theory to describe how most chemical reactions occur.
4. Describe how concentrations of reactants affect reaction rate.
5. Describe how temperature affects reaction rates.
6. Write equilibrium expressions for chemical reactions in solution.
7. Relate the magnitude of an equilibrium constant to the extent to which the forward or reverse reaction is favored.
8. Write equilibrium expressions for chemical reactions involving solids or liquids.
9. Calculate equilibrium constants from equilibrium concentrations of reactants and products.
10. Use equilibrium constants to calculate the equilibrium concentration of one of the reactants or products, given the equilibrium concentrations of the others.
11. State Le Châtelier's principle.
12. Use Le Châtelier's principle to predict the effect of a concentration change on an equilibrium.
13. Use Le Châtelier's principle to predict the effect of a volume change on an equilibrium.
14. Use Le Châtelier's principle to predict the effect of a temperature change on equilibrium.
15. Write solubility product constant expressions.
16. Define molar solubility.
17. Calculate molar solubility from K_{sp}.
18. Describe how activation energies affect reaction rates.
19. Describe how catalysts affect reaction rates.

Chapter 15

CHAPTER IN REVIEW

- The rate of a chemical reaction is the amount of reactant that changes to product in a given period of time.

- According to collision theory, most chemical reactions occur through collisions between molecules or atoms.

- The rate of a chemical reaction generally increases with increasing concentration of the reactants.

- The rate of a chemical reaction generally increases with increasing temperature of the reaction mixture.

- Reaction rates generally decrease as a reaction proceeds.

- When a chemical reaction is at dynamic equilibrium, the rate of the forward reaction equals the rate of the reverse reaction.

- The equilibrium constant (K_{eq}) for a reaction is defined as the ratio, at equilibrium, of the concentrations of the products raised to their stoichiometric coefficients divided by the concentrations of the reactants raised to their stoichiometric coefficients.

- The concentrations of pure solids and pure liquids are not included in equilibrium constant expressions.

- Equilibrium constants are temperature dependent.

- Le Châtelier's principle states that when a chemical system at equilibrium is disturbed, the system shifts in a direction that minimizes the disturbance.

- When a reaction "shifts to the right" it proceeds in the forward direction, consuming reactants and forming products. When a reaction "shifts to the left" it proceeds in the reverse direction, consuming products and forming reactants.

- If a chemical system is at equilibrium, increasing the concentration of one or more reactants causes the reaction to shift to the right.

- If a chemical system is at equilibrium, increasing the concentration of one or more products causes the reaction to shift to the left.

- If a chemical system is at equilibrium, decreasing the volume causes the reaction to shift in the direction that has the fewer number of moles of gas particles.

Chemical Equilibrium

- If a chemical system is at equilibrium, increasing the volume causes the reaction to shift in the direction that has the greater number of moles of gas particles.

- Endothermic reactions absorb heat. Exothermic reactions emit heat.

- In an endothermic chemical reaction, increasing the temperature causes the reaction to shift to the right; decreasing the temperature causes the reaction to shift to the left.

- In an exothermic chemical reaction, increasing the temperature causes the reaction to shift to the left; decreasing the temperature causes the reaction to shift to the right.

- The equilibrium expression for a chemical equation that represents the dissolving of an ionic compound is called the solubility-product constant (K_{sp}).

- The molar solubility of a compound is its solubility in moles per liter.

- The molar solubility of compound can be computed directly from K_{sp}.

- The activation energy of a chemical reaction is an energy barrier that normally exists between the reactants and the products.

- At a given temperature, the higher the activation energy is for a chemical reaction, the slower the reaction rate.

- A catalyst is a substance that increases the rate of a chemical reaction but is not consumed in the reaction. Catalysts work by lowering the activation energy for a reaction.

- Enzymes are biological catalysts that increase rates of biochemical reactions.

SKILLBUILDER PROBLEMS AND SOLUTIONS

SKILLBUILDER 15.1   Writing Equilibrium Expressions for Chemical Reactions

Write an equilibrium expression for the following chemical equation.

$$H_2(g) + F_2(g) \rightleftharpoons 2\,HF(g)$$

Solution:
The equilibrium expression is the concentration of the product raised to its stoichiometric coefficient divided by the concentrations of the reactants raised to their stoichiometric coefficients.

Chapter 15

$$K_{eq} = \frac{[HF]^2}{[H_2][F_2]}$$

SKILLBUILDER 15.2   **Writing Equilibrium Expressions for Reactions Involving a Solid or a Liquid**

Write an equilibrium expression for the following chemical equation.

$$4\ HCl(g) + O_2(g) \rightleftharpoons 2\ H_2O(l) + 2\ Cl_2(g)$$

Solution:
Since $H_2O(l)$ is a liquid, it is omitted from the equilibrium expression.

$$K_{eq} = \frac{[Cl_2]^2}{[HCl]^4[O_2]}$$

SKILLBUILDER 15.3   **Calculating Equilibrium Constants**

Consider the following reaction.

$$CO(g) + 2\ H_2(g) \rightleftharpoons CH_3OH(g)$$

A mixture of CO, H_2, and CH_3OH is allowed to come to equilibrium at 225 °C. The measured equilibrium concentrations are [CO] = 0.489 M, [H_2] = 0.146 M, and [CH_3OH] = 0.151 M. What is the value of the equilibrium constant at this temperature?

Given: [CO] = 0.489 M
       [H_2] = 0.146 M
       [CH_3OH] = 0.151 M

Find: K_{eq}

Solution:
The expression for K_{eq} is written from the balanced equation.

$$K_{eq} = \frac{[CH_3OH]}{[CO][H_2]^2}$$

To calculate the value of K_{eq}, simply substitute the correct equilibrium concentrations into the expression for K_{eq}.

$$K_{eq} = \frac{[0.151]}{[0.489][0.146]^2} = 14.5$$

SKILLBUILDER PLUS

Suppose that the preceding reaction is carried out at a different temperature and that the initial concentrations of the reactants are [CO] = 0.500 M and [H$_2$] = 1.00 M. Assuming that there is no product at the beginning of the reaction, and that at equilibrium [CO] = 0.15 M, find the equilibrium constant at this new temperature. Hint: Use the stoichiometric relationships from the balanced equation to find the equilibrium concentrations of H$_2$ and CH$_3$OH.

Given: Initial [CO] = 0.500 M
Initial [H$_2$] = 1.00 M
Equilibrium [CO] = 0.15 M

Balanced Equation:

$$CO(g) + 2\,H_2(g) \rightleftharpoons CH_3OH(g)$$

Conversion Factors:
1 mol CO ≡ 2 mol H$_2$
1 mol CO ≡ 1 mol CH$_3$OH

Find: K_{eq}

Solution Map:

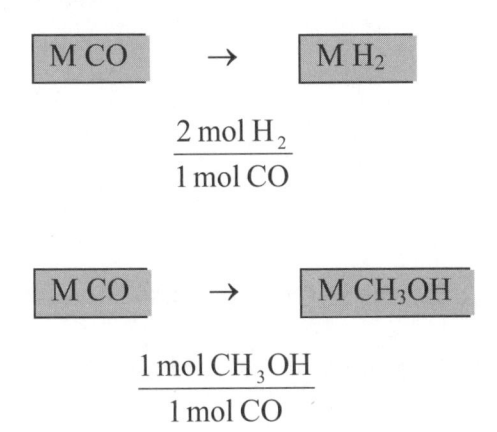

M CO → M H$_2$

$$\frac{2\text{ mol H}_2}{1\text{ mol CO}}$$

M CO → M CH$_3$OH

$$\frac{1\text{ mol CH}_3\text{OH}}{1\text{ mol CO}}$$

Solution:
The difference between the initial concentration of CO and the equilibrium concentration of CO is the concentration of CO that reacts.

0.500 M CO − 0.15 M CO = 0.35 M CO

Chapter 15

We calculate the concentration of H_2 that reacts using the stoichiometric relationship from the balanced equation.

$$\frac{0.35 \text{ mol CO}}{L} \times \frac{2 \text{ mol H}_2}{1 \text{ mol CO}} = 0.70 \frac{\text{mol H}_2}{L}$$

Since the initial concentration of H_2 is 1.00 M, and 0.70 M H_2 reacts, the equilibrium concentration of H_2 is

$$1.00 \text{ M H}_2 - 0.70 \text{ M H}_2 = 0.30 \text{ M H}_2$$

Next we calculate the concentration of CH_3OH that forms using the stoichiometric relationship from the balanced equation.

$$\frac{0.35 \text{ mol CO}}{L} \times \frac{1 \text{ mol CH}_3\text{OH}}{1 \text{ mol CO}} = 0.35 \frac{\text{mol CH}_3\text{OH}}{L}$$

Since there is no CO at the beginning of the reaction, the equilibrium concentration of CO is 0.35 M.

The expression for K_{eq} is written from the balanced equation.

$$K_{eq} = \frac{[CH_3OH]}{[CO][H_2]^2}$$

To calculate the value of K_{eq}, we substitute the equilibrium concentrations into the expression.

$$K_{eq} = \frac{[0.35]}{[0.15][0.30]^2}$$

$$= 26$$

SKILLBUILDER 15.4 Using Equilibrium Constants in Calculations

Diatomic iodine (I_2) decomposes at high temperature to form I atoms according to the following reaction.

$$I_2(g) \rightleftharpoons 2 \text{ I}(g) \qquad K_{eq} = 0.011 \text{ at } 1200 \text{ °C}$$

In an equilibrium mixture, the concentration of I_2 is 0.10 M. What is the equilibrium concentration of I?

Given: $[I_2] = 0.10$ M
$K_{eq} = 0.011$

Chemical Equilibrium

Find: [I]

Solution Map:

$$[I_2], K_{eq} \rightarrow [I]$$

$$K_{eq} = \frac{[I]^2}{[I_2]}$$

Solution:
We first solve the equilibrium expression for [I] and then substitute in the appropriate values to compute it.

$$K_{eq} = \frac{[I]^2}{[I_2]}$$

$$[I]^2 = K_{eq}[I_2]$$

$$[I] = \sqrt{K_{eq}[I_2]}$$

$$= \sqrt{0.011\,[0.10]}$$

$$= 0.033\ M$$

SKILLBUILDER 15.5   The Effect of a Concentration Change on Equilibrium

Consider the following reaction in chemical equilibrium.

$$2\ BrNO(g) \rightleftharpoons 2\ NO(g) + Br_2(g)$$

What is the effect of adding additional Br_2 to the reaction mixture? What is the effect of adding additional BrNO?

Solution:
Adding additional Br_2 increases the concentration of Br_2 and causes the reaction to shift to the left. Adding additional BrNO increases the concentration of BrNO and causes the reaction to shift to the right.

Chapter 15

SKILLBUILDER PLUS

What is the effect of removing some Br₂ from the preceding reaction mixture?

Solution:
Removing some Br₂ decreases the concentration of Br₂ and causes the reaction to shift to the right.

SKILLBUILDER 15.6   **The Effect of a Volume Change on Equilibrium**

Consider the following reaction at chemical equilibrium.

$$2\,SO_2(g) + O_2(g) \rightleftharpoons 2\,SO_3(g)$$

What is the effect of decreasing the volume of the reaction mixture? Increasing the volume of the reaction mixture?

Solution:
The chemical equation has 3 mol of gas on the left and 2 mol of gas on the right. Decreasing the volume of the reaction mixture increases the pressure and causes the reaction to shift to the right (toward the side with fewer moles of gas particles). Increasing the volume of the reaction mixture decreases the pressure and causes the reaction to shift to the left (toward the side with more moles of gas particles).

SKILLBUILDER 15.7   **The Effect of a Temperature Change on Equilibrium**

The following reaction is exothermic.

$$2\,SO_2(g) + O_2(g) \rightleftharpoons 2\,SO_3(g)$$

What is the effect of increasing the temperature of the reaction mixture? Decreasing the temperature?

Solution:
Since the reaction is exothermic, we can think of heat as a product.

$$2SO_2(g) + O_2(g) \rightleftharpoons 2SO_3(g) + Heat$$

Raising the temperature is adding heat, causing the reaction to shift to the left. Lowering the temperature is removing heat, causing the reaction to shift to the right.

Chemical Equilibrium

SKILLBUILDER 15.8 **Writing Expressions for K_{sp}**

Write expressions for K_{sp} for each of the following ionic compounds.

a) AgI
b) Ca(OH)$_2$

Solution:
To write the expression for K_{sp}, first write the chemical reaction showing the solid compound in equilibrium with its dissolved aqueous ions. Then write the equilibrium expression based on this equation.

a) $AgI(s) \rightleftharpoons Ag^+(aq) + I^-(aq)$

$$K_{sp} = [Ag^+][I^-]$$

b) $Ca(OH)_2(s) \rightleftharpoons Ca^{2+}(aq) + 2\,OH^-(aq)$

$$K_{sp} = [Ca^{2+}][OH^-]^2$$

SKILLBUILDER 15.9 **Calculating Molar Solubility from K_{sp}**

Calculate the molar solubility of CaSO$_4$.

Solution:
Begin by writing the reaction by which solid CaSO$_4$ dissolves into its constituent aqueous ions.

$$CaSO_4(s) \rightleftharpoons Ca^{2+}(aq) + SO_4^{2-}(aq)$$

Next, write the expression for K_{sp}.

$$K_{sp} = [Ca^{2+}][SO_4^{2-}]$$

Since [Ca^{2+}] or [SO$_4^{2-}$] at equilibrium equals the molar solubility (S), you can write

$$S = [Ca^{2+}] = [SO_4^{2-}]$$

Substituting this into the equilibrium expression.

$$K_{sp} = S \times S = S^2$$

Therefore,

Chapter 15

$$S = \sqrt{K_{sp}}$$

Finally, look up the value of K_{sp} in Table 15.2 in the text and compute S.

$$\begin{aligned} S &= \sqrt{K_{sp}} \\ &= \sqrt{7.10 \times 10^{-5}} \\ &= 8.43 \times 10^{-3} \end{aligned}$$

The molar solubility of $CaSO_4$ is 8.43×10^{-3} moles per liter.

SELF-TEST QUESTIONS

A. Match the following terms with the phrases below.

> activation energy
> collision theory
> dynamic equilibrium
> equilibrium constant (K_{eq})
> Le Châtelier's principle
> molar solubility
> rate of a chemical reaction (reaction rate)
> solubility-product constant (K_{sp})

1. Amount of reactant that changes to product in a given period of time
2. Theory that says that most chemical reactions occur through collisions between molecules or atoms
3. Energy barrier for a chemical reaction
4. Condition in which the rate of the forward reaction is equal to the rate of the reverse reaction
5. The ratio, at equilibrium, of the concentrations of the products of the reaction raised to their stoichiometric coefficients divided by the concentrations of the reactants raised to their stoichiometric coefficients
6. When a chemical system at equilibrium is disturbed, the system shifts in a direction that minimizes the disturbance
7. Equilibrium expression for a chemical equation that represents the dissolving of an ionic compound
8. Solubility expressed in units of moles per liter

B. True/False

1. All collisions between reactants lead to products.
2. As a chemical reaction proceeds, the concentration of reactants increases.
3. Reaction rates generally increase with increasing temperature.
4. At dynamic equilibrium, the forward and reverse reaction rates are equal.
5. Equilibria in which the products are favored have high equilibrium constants.
6. Pure liquids and solids are excluded from equilibrium expressions.

7. If a chemical system is at equilibrium, an increase in the concentration of a reactant causes the reaction to shift to the right.
8. In an exothermic chemical reaction, increasing the temperature causes the reaction to shift to the right.
9. Molar solubility is symbolized by K_{sp}.
10. Enzymes increase the activation energy of biochemical reactions.

C. Multiple Choice

1. Consider the following reaction.

$$H_2(g) + Cl_2(g) \rightarrow 2\ HCl(g)$$

   Which of the following is expected to cause the reaction rate to increase?
   a) An increase in concentration of H_2
   b) A decrease in concentration of Cl_2
   c) A decrease in concentration of H_2
   d) An increase in concentration of HCl

2. What is the equilibrium expression for the following reaction?

$$N_2(g) + 3\ H_2(g) \rightleftharpoons 2\ NH_3(g)$$

   a) $K_{eq} = \dfrac{[NH_3]}{[N_2][H_2]}$

   b) $K_{eq} = \dfrac{[NH_3]^2}{[N_2][H_2]}$

   c) $K_{eq} = \dfrac{[NH_3]}{[N_2][H_2]^3}$

   d) $K_{eq} = \dfrac{[NH_3]^2}{[N_2][H_2]^3}$

Chapter 15

3. What is the equilibrium expression for the following heterogeneous reaction?

$$2\ ZnS(s) + 3\ O_2(g) \rightleftharpoons 2\ ZnO(s) + 2\ SO_2(g)$$

a) $K_{eq} = \dfrac{[ZnO][SO_2]}{[ZnS][O_2]}$

b) $K_{eq} = \dfrac{[ZnO]^2[SO_2]^2}{[ZnS]^2[O_2]^3}$

c) $K_{eq} = \dfrac{[SO_2]^2}{[ZnS]^2[O_2]^3}$

d) $K_{eq} = \dfrac{[SO_2]^2}{[O_2]^3}$

4. Consider the reaction shown below at 25 °C.

$$CO(g) + Cl_2(g) \rightleftharpoons COCl_2(g)$$

The equilibrium concentrations are [CO] = 2.4 × 10⁻⁶ M, [Cl₂] = 1.2 × 10⁻⁶ M, and [COCl₂] = 8.4 × 10⁻² M. Calculate K_{eq} at 25 °C.

a) 3.4 × 10⁻¹¹
b) 8.4 × 10⁻²
c) 3.4 × 10⁻¹
d) 2.9 × 10¹⁰

5. Consider the reaction shown below at 127 °C.

$$PCl_3(g) + Cl_2(g) \rightleftharpoons PCl_5(g) \qquad K_{eq} = 96$$

At equilibrium, the concentration of PCl₃ is 3.6 × 10⁻² M and the concentration of Cl₂ is 4.6 × 10⁻³ M. Calculate the equilibrium concentration of PCl₅.

a) 1.3 × 10⁻³ M
b) 8.2 × 10⁻² M
c) 1.6 × 10⁻² M
d) 1.2 × 10¹ M

Chemical Equilibrium

6. Consider the following reaction at equilibrium.

$$F_2(g) + Cl_2(g) \rightleftharpoons 2\, ClF(g)$$

Which of the following would cause the reaction to shift to the right?
a) Removal of $F_2(g)$
b) Removal of $ClF(g)$
c) Removal of $Cl_2(g)$
d) Addition of $ClF(g)$

7. Consider the following reaction at equilibrium.

$$I_2(g) \rightleftharpoons 2\, I(g)$$

Which of the following would cause the reaction to shift to the left?
a) A decrease in the volume of the container
b) An increase in the volume of the container
c) Addition of $I_2(g)$
d) Removal of $I(g)$

8. The reaction shown below is endothermic.

$$2\, NOCl(g) \rightleftharpoons 2\, NO(g) + Cl_2(g)$$

Which of the following would cause the reaction to shift to the right?
a) A decrease in the volume of the container
b) An increase in temperature
c) Addition of $NO(g)$
d) Removal of $NOCl(g)$

9. Write the K_{sp} expression for Ag_2CrO_4.
a) $K_{sp} = \dfrac{[Ag^+][CrO_4^{2-}]}{[Ag_2CrO_4]}$
b) $K_{sp} = \dfrac{[Ag^+]^2[CrO_4^{2-}]}{[Ag_2CrO_4]}$
c) $K_{sp} = [Ag^+][CrO_4^{2-}]$
d) $K_{sp} = [Ag^+]^2[CrO_4^{2-}]$

10. Calculate the molar solubility of $PbSO_4$. $K_{sp} = 1.82 \times 10^{-8}$.
a) 3.31×10^{-16} M
b) 1.35×10^{-8} M
c) 1.82×10^{-8} M
d) 1.35×10^{-4} M

Chapter 15

D. Crossword Puzzle

ACROSS

1. Reaction that emits heat
3. Water that contains calcium carbonate and magnesium carbonate
5. Substance that increases the rate of a chemical reaction
6. Reaction that absorbs heat

DOWN

2. Biological catalyst
4. Reaction that proceeds in both the forward and reverse directions

ANSWERS TO SELF-TEST QUESTIONS

A. Matching
1. rate of a chemical reaction (reaction rate) 2. collision theory 3. activation energy
4. dynamic equilibrium 5. equilibrium constant (K_{eq}) 6. Le Châtelier's principle
7. solubility-product constant (K_{sp}) 8. molar solubility

B. True/False
1. F 2. F 3. T 4. T 5. T 6. T 7. T 8. F 9. F 10. F

C. Multiple Choice
1. a 2. d 3. d 4. d 5. c 6. b 7. a 8. b 9. d 10. d

Chemical Equilibrium

D. Crossword Puzzle

¹E	X	O	T	H	²E	R	M	I	C		
					N				³H	A	⁴R	D
					Z						E	
⁵C	A	T	A	L	Y	S	T				V	
					M						E	
				⁶E	N	D	O	T	H	E	R	M	I	C
											S	
											I	
											B	
											L	
											E	

Chapter 15

Oxidation and Reduction

16

CHAPTER OVERVIEW

Chapter 16 focuses on oxidation and reduction. Oxidation number assignments and steps for balancing redox reactions are presented. The activity series is used to predict spontaneous redox reactions. Chemical reactions that generate electricity in batteries are described. Electrolysis and corrosion are discussed.

CHAPTER OBJECTIVES

After reading and studying the text, students should be able to:

1. Differentiate between oxidation and reduction.
2. Differentiate between oxidizing agents and reducing agents.
3. Assign oxidation numbers.
4. Use oxidation states to identify oxidation and reduction.
5. Balance redox equations.
6. Use the activity series to predict spontaneous redox reactions.
7. Use the activity series to predict whether a metal will dissolve in acid.
8. Describe the components of an electrochemical cell.
9. Describe the common dry cell battery.
10. Describe the lead-acid storage battery.
11. Describe a fuel cell.
12. Differentiate between a galvanic cell and an electrolytic cell.
13. Give common examples of corrosion.
14. List ways to prevent corrosion.

CHAPTER IN REVIEW

- Oxidation-reduction, or redox, reactions are reactions that involve the transfer of electrons.

- Oxidation is the loss of electrons, reduction is the gain of electrons.

- An oxidizing agent is the substance being reduced in a redox reaction. A reducing agent is the substance being oxidized in a redox reaction.

Chapter 16

- When a substance is oxidized, its oxidizing state increases. When a substance is reduced, its oxidation state decreases.

- Oxidation numbers, or oxidation states, are assigned to each element in an oxidation-reduction reaction for the purpose of keeping track of electrons. The oxidation state of an atom in a free element is 0. The oxidation state of a monoatomic ion is equal to its charge. The sum of the oxidations states of all atoms in a molecule or formula unit is 0. The sum of the oxidation states of all atoms in a polyatomic ion is equal to the charge of the ion.

- Balance redox equations using the half-reaction method. 1) Assign oxidation states to all atoms and identify the substances being oxidized and reduced. 2) Separate the overall reaction into two half-reactions. 3) Balance each half-reaction by first balancing all elements other than hydrogen and oxygen. Balance oxygen by adding water. Balance hydrogen by adding H^+. 4) Balance each half-reaction with respect to charge by adding electrons. 5) Make the number of electrons in both half-reactions equal by multiplying one or both half-reactions by a small whole number. 6) Add the two half-reactions together, canceling electrons and other species as necessary. 7) Verify that the reaction is balanced both with respect to mass and with respect to charge.

- The activity series of metals lists metals in order of decreasing tendency to lose electrons. Any half-reaction on the list will be spontaneous when paired with the reverse of any half-reaction below it.

- Metals above H_2 on the activity series will dissolve in acids, metals below H_2 will not dissolve in acids.

- Electrical current is the flow of electronic charge.

- In an electrochemical cell, the anode is the electrode where oxidation occurs. The cathode is the electrode where reduction occurs.

- An electrochemical cell that spontaneously produces electrical current is called a galvanic or voltaic cell.

- Dry cells are batteries that do not contain large amounts of water. Alkaline batteries contain a base. Most automobile batteries are lead-acid storage batteries.

- In fuel cells the reactants (fuel) constantly flow through the battery, generating electrical current as they undergo a redox reaction.

- Electrolysis is a process in which electrical current is used to drive an otherwise nonspontaneous redox reaction. Electrochemical cells used for electrolysis are called electrolytic cells.

- Corrosion is the oxidation of metals.

SKILLBUILDER PROBLEMS AND SOLUTIONS

SKILLBUILDER 16.1 Identifying Oxidation and Reduction

For each of the following reactions, identify the substance being oxidized and the substance being reduced.
a) $2 K(s) + Cl_2(g) \rightarrow 2 KCl(s)$
b) $2 Al(s) + 3 Sn^{2+}(aq) \rightarrow 2 Al^{3+}(aq) + 3 Sn(s)$
c) $C(s) + O_2(g) \rightarrow CO_2(g)$

Solution:
a) $2 K(s) + Cl_2(g) \rightarrow 2 KCl(s)$
In this reaction, a metal (potassium) is reacting with an electronegative nonmetal (Cl_2). K loses electrons and is therefore oxidized, while Cl_2 gains electrons and is therefore reduced.

b) $2 Al(s) + 3 Sn^{2+}(aq) \rightarrow 2 Al^{3+}(aq) + 3 Sn(s)$
In this reaction electrons are transferred from Al to Sn^{2+}. Al loses electrons and is oxidized. Sn^{2+} gains electrons and is reduced.

c) $C(s) + O_2(g) \rightarrow CO_2(g)$
In this reaction, carbon is gaining oxygen and losing electrons to oxygen. Carbon is therefore oxidized and O_2 is reduced.

SKILLBUILDER 16.2 Identifying Oxidizing and Reducing Agents

For each of the following reactions, identify the oxidizing agent and the reducing agent.
a) $2 K(s) + Cl_2(g) \rightarrow 2 KCl(s)$
b) $2 Al(s) + 3 Sn^{2+}(aq) \rightarrow 2 Al^{3+}(aq) + 3 Sn(s)$
c) $C(s) + O_2(g) \rightarrow CO_2(g)$

Solution:
In the previous skillbuilder, we identified the substance being oxidized and reduced for each of these reactions. The substance being oxidized is the reducing agent and the substance being reduced is the oxidizing agent.

a) K is oxidized and is therefore the reducing agent; Cl_2 is reduced and is therefore the oxidizing agent.

b) Al is oxidized and is therefore the reducing agent; Sn^{2+} is reduced and is therefore the oxidizing agent.

Chapter 16

c) C is oxidized and is therefore the reducing agent; O_2 is reduced and is therefore the oxidizing agent.

SKILLBUILDER 16.3   **Assigning Oxidation States**

Assign an oxidation state to each atom in the following.

a) Zn
b) Cu^{2+}
c) $CaCl_2$
d) CF_4
e) NO_2^-
f) SO_3

Solution:
a) Zn
   Since Zn is a free element, the oxidation state of Zn is 0 (Rule 1).

   Zn
   0

b) Cu^{2+}
   Since Cu^{2+} is a monoatomic ion, the oxidation state of Cu^{2+} is +2 (Rule 2).

   Cu^{2+}
   +2

c) $CaCl_2$
   The oxidation state of Ca is +2. The oxidation state of Cl is −1. Since this is a neutral compound, the sum of the oxidation states is 0 (Rule 3).

   Ca Cl_2
   +2 −1
   sum: +2 + 2(−1) = 0

d) CF_4
   The oxidation state of fluorine is −1 (Rule 5). The oxidation state of carbon must be deduced from Rule 3, which states that the sum of the oxidation states of all atoms in a neutral molecule must be 0. Since there are four fluorine atoms, the oxidation state of F must be multiplied by 4 when computing the sum.

   (C ox state) + 4(F ox state) = 0
   (C ox state) + 4(−1) = 0
   (C ox state) − 4 = 0
   C ox state = +4

Oxidation and Reduction

C F$_4$
+4 −1
sum: +4 + 4(−1) = 0

e) NO$_2^-$

The oxidation state of O is −2 (Rule 5). The oxidation state of N is expected to be −3 (Rule 5). However, if that were the case, the sum of the oxidation states would not equal the charge of the ion. Since O is higher on the hierarchical list, it takes priority and the oxidation state of nitrogen is computed by setting the sum of all of the oxidation states equal to −1, the charge of the ion (Rule 3).

(N ox state) + 2(O ox state) = −1
(N ox state) + 2(−2) = −1 (oxygen takes priority over nitrogen)
(N ox state) − 4 = −1
N ox state = −1 + 4
N ox state = +3

N O$_2^-$
+3 −2
sum: +3 + 2(−2) = −1

f) SO$_3$

The oxidation state of O is −2 (Rule 5). The oxidation state of sulfur must be deduced from Rule 3, which states that the sum of the oxidation states of all atoms in a neutral molecule must be 0. Since there are three oxygen atoms, the oxidation state of O must be multiplied by 3 when computing the sum.

(S ox state) + 3(O ox state) = 0
(S ox state) + 3(−2) = 0
(S ox state) − 6 = 0
S ox state = +6

S O$_3$
+6 −2
sum: +6 + 3(−2) = 0

SKILLBUILDER 16.4 Using Oxidation States to Identify Oxidation and Reduction

Use oxidation states to identify the element that is being oxidized and the element that is being reduced in the following redox reaction.

$$Sn(s) + 4\ HNO_3(aq) \rightarrow SnO_2(s) + 4\ NO_2(g) + 2\ H_2O(g)$$

Solution:
Begin by assigning oxidation states to each atom in the reaction.

Chapter 16

$$Sn(s) + 4HNO_3(aq) \rightarrow SnO_2(s) + 4NO_2(g) + 2H_2O(g)$$
$$0 +1\ +5\ -2 +4\ -2 +4\ -2 +1\ -2$$

Since Sn increases in oxidation state, it is oxidized. Since N decreases in oxidation state, it is reduced.

SKILLBUILDER 16.5   **Balancing Redox Equations Using the Half-Reaction Method**

Balance the following redox reaction occurring in acidic solution.

$$H^+(aq) + Cr(s) \rightarrow H_2(g) + Cr^{3+}(aq)$$

Solution:
Assign oxidation numbers.

$$H^+(aq) + Cr(s) \rightarrow H_2(g) + Cr^{3+}(aq)$$
$$+1 0 0 \phantom{\ \ \ \ Cr^{3+}(aq)}+3$$

Oxidation: $Cr(s) \rightarrow Cr^{3+}(aq)$
Reduction: $H^+(aq) \rightarrow H_2(g)$

Chromium is balanced; balance hydrogen.

$$Cr(s) \rightarrow Cr^{3+}(aq)$$
$$2\,H^+(aq) \rightarrow H_2(g)$$

Balance each half-reaction with respect to charge by adding electrons.

$$Cr(s) \rightarrow Cr^{3+}(aq) + 3\,e^-$$
$$2\,e^- + 2\,H^+(aq) \rightarrow H_2(g)$$

Make the number of electrons in both half-reactions equal by multiplying one or both half-reactions by a small whole number.

$$2 \times [Cr(s) \rightarrow Cr^{3+}(aq) + 3\,e^-]$$
$$3 \times [2\,e^- + 2\,H^+(aq) \rightarrow H_2(g)]$$

Add the two half-reactions together, canceling electrons and other species as necessary.

$$2\,Cr(s) \rightarrow 2\,Cr^{3+}(aq) + 6\,e^-$$
$$\underline{6\,e^- + 6\,H^+(aq) \rightarrow 3\,H_2(g)}$$
$$2\,Cr(s) + 6\,H^+(aq) \rightarrow 2\,Cr^{3+}(aq) + 3\,H_2(g)$$

Oxidation and Reduction

SKILLBUILDER 16.6 **Balancing Redox Equations Using the Half-Reaction Method**

Balance the following redox reaction occurring in acidic solution.

$$Cu(s) + NO_3^-(aq) \rightarrow Cu^{2+}(aq) + NO_2(g)$$

Solution:
Assign oxidation numbers.

$$Cu(s) + NO_3^-(aq) \rightarrow Cu^{2+}(aq) + NO_2(g)$$
$$0 +5\,-2 \phantom{Cu^{2+}}+2 +4\,-2$$

Oxidation: $Cu(s) \rightarrow Cu^{2+}(aq)$
Reduction: $NO_3^-(aq) \rightarrow NO_2(g)$

All elements other than oxygen are balanced. Balance oxygen by adding H_2O.

$$Cu(s) \rightarrow Cu^{2+}(aq)$$
$$NO_3^-(aq) \rightarrow NO_2(g) + \mathbf{H_2O}(l)$$

Balance hydrogen by adding H^+.

$$Cu(s) \rightarrow Cu^{2+}(aq)$$
$$\mathbf{2\,H^+}(aq) + NO_3^-(aq) \rightarrow NO_2(g) + H_2O(l)$$

Balance each half-reaction with respect to charge by adding electrons.

$$Cu(s) \rightarrow Cu^{2+}(aq) + \mathbf{2\,e^-}$$
$$\mathbf{1\,e^-} + 2\,H^+(aq) + NO_3^-(aq) \rightarrow NO_2(g) + H_2O(l)$$

Make the number of electrons in both half-reactions equal by multiplying one or both half-reactions by a small whole number.

$$Cu(s) \rightarrow Cu^{2+}(aq) + 2\,e^-$$
$$2 \times [1\,e^- + 2\,H^+(aq) + NO_3^-(aq) \rightarrow NO_2(g) + H_2O(l)]$$

Add the two half-reactions together, canceling electrons and other species as necessary.

$$Cu(s) \rightarrow Cu^{2+}(aq) + 2\,e^-$$
$$\underline{2\,e^- + 4\,H^+(aq) + 2\,NO_3^-(aq) \rightarrow 2\,NO_2(g) + 2\,H_2O(l)}$$
$$Cu(s) + 4\,H^+(aq) + 2\,NO_3^-(aq) \rightarrow Cu^{2+}(aq) + 2\,NO_2(g) + 2\,H_2O(l)$$

Chapter 16

SKILLBUILDER 16.7   **Balancing Redox Reactions**

Balance the following redox reaction occurring in acidic solution.

$$Sn(s) + MnO_4^-(aq) \rightarrow Sn^{2+}(aq) + Mn^{2+}(aq)$$

Solution:
Assign oxidation numbers.

$$Sn(s) + MnO_4^-(aq) \rightarrow Sn^{2+}(aq) + Mn^{2+}(aq)$$
$$0 +7\ -2 +2 +2$$

Oxidation: $Sn(s) \rightarrow Sn^{2+}(aq)$
Reduction: $MnO_4^-(aq) \rightarrow Mn^{2+}(aq)$

All elements other than oxygen are balanced. Balance oxygen by adding H_2O.

$$Sn(s) \rightarrow Sn^{2+}(aq)$$
$$MnO_4^-(aq) \rightarrow Mn^{2+}(aq) + \mathbf{4\ H_2O}(l)$$

Balance hydrogen by adding H^+.

$$Sn(s) \rightarrow Sn^{2+}(aq)$$
$$\mathbf{8\ H^+}(aq) + MnO_4^-(aq) \rightarrow Mn^{2+}(aq) + 4\ H_2O(l)$$

Balance each half-reaction with respect to charge by adding electrons.

$$Sn(s) \rightarrow Sn^{2+}(aq) + \mathbf{2\ e^-}$$
$$\mathbf{5\ e^-} + 8\ H^+(aq) + MnO_4^-(aq) \rightarrow Mn^{2+}(aq) + 4\ H_2O(l)$$

Make the number of electrons in both half-reactions equal by multiplying one or both half-reactions by a small whole number.

$$\mathbf{5} \times [Sn(s) \rightarrow Sn^{2+}(aq) + 2\ e^-]$$
$$\mathbf{2} \times [5\ e^- + 8\ H^+ + MnO_4^-(aq) \rightarrow Mn^{2+}(aq) + 4\ H_2O(l)]$$

Add the two half-reactions together, canceling electrons and other species as necessary.

$$5\ Sn(s) \rightarrow 5\ Sn^{2+}(aq) + 10\ e^-$$
$$\underline{10\ e^- + 16\ H^+(aq) + 2\ MnO_4^-(aq) \rightarrow 2\ Mn^{2+}(aq) + 8\ H_2O(l)}$$
$$5\ Sn(s) + 16\ H^+(aq) + 2\ MnO_4^-(aq) \rightarrow 5\ Sn^{2+}(aq) + 2\ Mn^{2+}(aq) + 8\ H_2O(l)$$

Oxidation and Reduction

SKILLBUILDER 16.8 Predicting Spontaneous Redox Reactions

Will the following redox reactions be spontaneous?

a) $Zn(s) + Ni^{2+}(aq) \rightarrow Zn^{2+}(aq) + Ni(s)$
b) $Zn(s) + Ca^{2+}(aq) \rightarrow Zn^{2+}(aq) + Ca(s)$

Solution:
a) $Zn(s) + Ni^{2+}(aq) \rightarrow Zn^{2+}(aq) + Ni(s)$

   This reaction involves the oxidation of Zn

   $Zn(s) \rightarrow Zn^{2+}(aq) + 2\,e^-$

   with the reverse of a half-reaction *below it* in the activity series.

   $2\,e^- + Ni^{2+}(aq) \rightarrow Ni(s)$

   Therefore, the reaction *will be* spontaneous.

b) $Zn(s) + Ca^{2+}(aq) \rightarrow Zn^{2+}(aq) + Ca(s)$

   This reaction involves the oxidation of Zn

   $Zn(s) \rightarrow Zn^{2+}(aq) + 2\,e^-$

   with the reverse of a half-reaction *above it* in the activity series.

   $2\,e^- + Ca^{2+}(aq) \rightarrow Ca(s)$

   Therefore, the reaction *will not be* spontaneous.

SKILLBUILDER 16.9 Predicting Whether a Metal Will Dissolve in Acid

Will Ag dissolve in hydrobromic acid?

Solution:
No. Since Ag is below H_2 in the activity series, it will not dissolve in HBr.

Chapter 16

SELF-TEST QUESTIONS

A. Match the following terms with the phrases below.

    activity series of metals   half-reaction
    dry cell   lead-acid storage battery
    electrical current   oxidation state (oxidation number)
    electrochemical cell   oxidizing agent
    electrolytic cell   redox (oxidation-reduction) reaction
    fuel cell   reducing agent
    galvanic (voltaic) cell   salt bridge
    half-cell

1. Battery in which the fuel constantly flows through the battery generating electrical current
2. Reaction involving the transfer of electrons
3. Substance that is oxidized
4. Substance that is reduced
5. Number assigned to an atom in order to track electrons
6. Oxidation portion, or reduction portion, of an oxidation-reduction reaction
7. Table listing metals in order of decreasing tendency to lose electrons
8. Flow of electronic charge
9. Device that involves electrical current and a redox reaction
10. Portion of an electrochemical cell in which one of the half-reactions occurs
11. Inverted U-shaped tube containing a strong electrolyte that joins two half-cells
12. Electrochemical cell that spontaneously produces electrical current
13. Electrochemical cell that does not contain large amounts of liquid water
14. Battery found in most automobiles
15. Electrochemical cell used for electrolysis

B. True/False

1. Oxidation is loss of electrons.
2. In a redox reaction, the substance being reduced is the reducing agent.
3. The oxidation state of a monoatomic ion is 0.
4. In compounds, group 2 metals have an oxidation state of +2.
5. Metals above hydrogen on the activity series will dissolve in acids.
6. In a galvanic cell, the anode is labeled with a positive (+) sign.
7. The batteries found in most automobiles are dry cell batteries.
8. In electrolysis, an electrical current drives an otherwise nonspontaneous redox reaction.
9. The most common fuel cell is the nitrogen-oxygen fuel cell.
10. Corrosion is the reduction of metals.

C. Multiple Choice

1. Identify the species that is reduced in the following reaction.

$$2\ FeBr_3(aq) + 3\ Cl_2(g) \rightarrow 2\ FeCl_3(aq) + 3\ Br_2(g)$$

   a) FeBr$_3$
   b) Cl$_2$
   c) FeCl$_3$
   d) Br$_2$

2. Identify the reducing agent in the following reaction.

$$Zn(s) + CuSO_4(aq) \rightarrow ZnSO_4(aq) + Cu(s)$$

   a) Zn
   b) CuSO$_4$
   c) ZnSO$_4$
   d) Cu

3. Assign the oxidation number of C in C$_2$H$_5$OH.
   a) −2
   b) −4
   c) +2
   d) +4

4. Assign the oxidation number of Cr in Cr$_2$O$_7^{2-}$.
   a) +2
   b) +3
   c) +6
   d) +7

5. Balance the following redox reaction occurring in acidic aqueous solution.

$$Br_2(l) + SO_2(g) \rightarrow Br^-(aq) + SO_4^{2-}(aq)$$

   a) $Br_2(l) + SO_2(g) \rightarrow 2 Br^-(aq) + SO_4^{2-}(aq)$
   b) $2 H_2O(l) + Br_2(l) + SO_2(g) \rightarrow 2 Br^-(aq) + SO_4^{2-}(aq)$
   c) $2 H_2O(l) + Br_2(l) + SO_2(g) \rightarrow 2 Br^-(aq) + SO_4^{2-}(aq) + 2 H^+(aq)$
   d) $2 H_2O(l) + Br_2(l) + SO_2(g) \rightarrow 2 Br^-(aq) + SO_4^{2-}(aq) + 4 H^+(aq)$

Chapter 16

6. Balance the following redox reaction occurring in acidic aqueous solution.

$$I_2(aq) + S_2O_3^{2-}(aq) \rightarrow I^-(aq) + S_4O_6^{2-}(aq)$$

a) $I_2(aq) + S_2O_3^{2-}(aq) \rightarrow 2\,I^-(aq) + S_4O_6^{2-}(aq)$
b) $2\,I_2(aq) + 2\,S_2O_3^{2-}(aq) \rightarrow 4\,I^-(aq) + S_4O_6^{2-}(aq)$
c) $I_2(aq) + 2\,S_2O_3^{2-}(aq) \rightarrow I^-(aq) + S_4O_6^{2-}(aq)$
d) $I_2(aq) + 2\,S_2O_3^{2-}(aq) \rightarrow 2\,I^-(aq) + S_4O_6^{2-}(aq)$

7. Which of the following metals will spontaneously be oxidized when placed in an aqueous solution of Zn^{2+}?
a) Mg
b) Sn
c) Fe
d) Ni

8. Which of the following metals will dissolve in acid?
a) Cu
b) Ag
c) Au
d) Fe

9. An electrochemical cell has the following reaction occurring at the anode.

$$Mn(s) \rightarrow Mn^{2+}(aq) + 2\,e^-$$

Which of the following cathode reactions would produce a battery with the highest voltage?
a) $Fe^{2+}(aq) + 2\,e^- \rightarrow Fe(s)$
b) $Pb^{2+}(aq) + 2\,e^- \rightarrow Pb(s)$
c) $Sn^{2+}(aq) + 2\,e^- \rightarrow Sn(s)$
d) $Cu^{2+}(aq) + 2\,e^- \rightarrow Cu(s)$

10. Which of the following metals, if coated onto iron, would prevent the corrosion of iron?
a) Ni
b) Sn
c) Pb
d) Cr

D. Crossword Puzzle

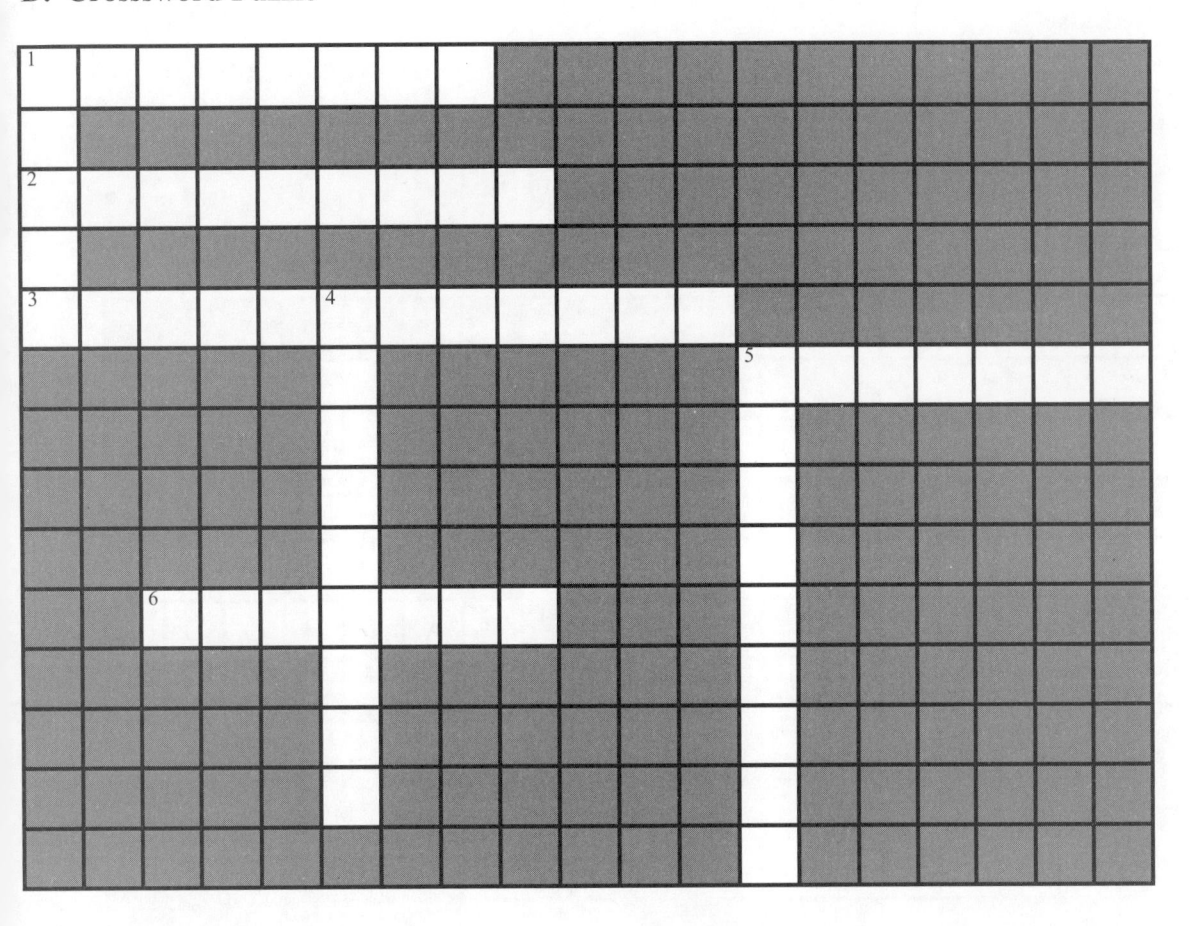

ACROSS

1. Battery that employs half-reactions that use a base
2. Loss of electrons
3. Process in which electrical current is used to drive an otherwise nonspontaneous redox reaction
5. Electrode where reduction occurs
6. Potential difference

DOWN

1. Electrode where oxidation occurs
4. Gain of electrons
5. Oxidation of metals

ANSWERS TO SELF-TEST QUESTIONS

A. Matching

1. fuel cell 2. redox (oxidation-reduction) reaction 3. reducing agent 4. oxidizing agent 5. oxidation state (oxidation number) 6. half-reaction 7. activity series of metals 8. electrical current 9. electrochemical cell 10. half-cell 11. salt bridge 12. galvanic (voltaic cell) 13. dry cell 14. lead-acid storage battery 15. electrolytic cell

Chapter 16

B. True/False
1. T 2. F 3. F 4. T 5. T 6. F 7. F 8. T 9. F 10. F

C. Multiple Choice
1. b 2. a 3. a 4. c 5. d 6. d 7. a 8. d 9. d 10. d

D. Crossword Puzzle

¹A	L	K	A	L	I	N	E						
N													
²O	X	I	D	A	T	I	O	N					
D													
³E	L	E	C	T	⁴R	O	L	Y	S	I	S		
					E					⁵C	A	T	H	O	D	E
					D					O			
					U					R			
					C					R			
			⁶V	O	L	T	A	G	E	O			
					I					S			
					O					I			
					N					O			
										N			

Radioactivity and Nuclear Chemistry   17

CHAPTER OVERVIEW

Chapter 17 explores radioactivity and nuclear chemistry. Radiation types and equations for nuclear reactions are presented. The concept of half-life is discussed. Radioactivity detectors are described. Fission and fusion are examined. Effects of radiation on living organisms are discussed.

CHAPTER OBJECTIVES

After reading and studying the text, students should be able to:

1. Describe the major contributions of Antoine-Henri Becquerel, Marie Sklodowska Curie, and Ernest Rutherford to nuclear chemistry.
2. Write isotopic symbols.
3. Write the symbol for an alpha particle.
4. Describe the characteristics of alpha particles.
5. Identify parent nuclides and daughter nuclides in nuclear reactions.
6. Write nuclear equations for alpha decay.
7. Write the symbol for a beta particle.
8. Describe the characteristics of beta particles.
9. Write nuclear equations for beta decay.
10. Write the symbol for a gamma ray.
11. Describe the characteristics of gamma radiation.
12. Write the symbol for a positron.
13. Describe the characteristics of positrons.
14. Write nuclear equations for positron emission.
15. List three types of radioactivity detectors.
16. Explain how a film-badge dosimeter works.
17. Explain how a Geiger-Müller counter works.
18. Explain how a scintillation counter works.
19. List sources of natural radioactivity.
20. Define half-life.
21. Use half-life to determine the amount of time required for a given amount of a radioactive sample to decay.
22. Explain what a radioactive decay series is.
23. Identify the primary source of radon in our environment.
24. Identify the source of carbon-14 in our environment.
25. Use carbon-14 content to determine the age of fossils or artifacts.

26. Describe the discoveries of Enrico Fermi, Lise Meitner, Fritz Strassmann, and Otto Hahn.
27. Explain what critical mass is.
28. State the main goal of J. R. Oppenheimer and the Manhattan Project.
29. Describe how nuclear fission can be used to generate electricity.
30. Explain the purpose of control rods in a nuclear reaction core.
31. List the main problems associated with nuclear electricity generation.
32. Compare nuclear fusion to nuclear fission.
33. Describe how radiation affects molecules in living systems.
34. Explain how radiation can increase the risk for cancer.
35. Explain how radiation can cause genetic defects.
36. Give the unit that is often used to report radiation exposure.
37. Describe the outcomes of radiation exposure at different doses.
38. Explain the role of isotope scanning in medicine.
39. Explain how radioactivity is used to treat cancer.

CHAPTER IN REVIEW

- Radioactivity is the emission of tiny, invisible particles by the nuclei of certain atoms.

- Atoms that emit radioactivity are radioactive.

- Antoine-Henri Becquerel discovered radioactivity.

- Phosphorescence is the long-lived emission of light that sometimes follows the absorption of light by some atoms and molecules.

- Alpha (α) radiation occurs when an unstable nucleus emits a small particle composed of two protons and two neutrons. The symbol for an alpha particle is identical to the symbol for helium-4.

- A nuclear equation is an equation that represents nuclear processes such as radioactivity. The original atom is called the parent nuclide; the products are called daughter nuclides.

- The sum of the atomic numbers on both sides of a nuclear equation must be equal and the sum of the mass numbers on both sides must be equal.

- Ionizing power is the ability of radiation to ionize other molecules and atoms.

- Penetrating power is the ability to penetrate matter.

- Alpha particles have a high ionizing power and a low penetrating power.

- Beta (β) radiation occurs when an unstable nucleus emits an electron.

- Beta particles have intermediate ionizing power and intermediate penetrating power.

Radioactivity and Nuclear Chemistry

- Gamma (γ) radiation is electromagnetic radiation—high energy, short wavelength photons.

- Gamma rays have low ionizing power and high penetrating power.

- Positron emission occurs when a proton in an unstable nucleus changes into a neutron and emits a positron.

- A positron has the mass of an electron but has a +1 charge.

- Film-badge dosimeters are radiation detectors which consist of photographic film that becomes exposed when radiation passes through.

- Geiger-Müller counters are radiation detectors in which radioactive particles pass through an argon-filled chamber, creating a trail of ionized argon atoms. The ions produce an electrical signal that can be detected on a meter or as an audible click.

- Scintillation counters are radiation detectors in which radioactive particles pass through a salt that emits ultraviolet or visible light in response to excitation by radioactive particles.

- Half-life is the amount of time required for one-half of the parent nuclides in a radioactive sample to decay.

- Radiocarbon dating is a technique used to determine the age of fossils and artifacts based on carbon-14 concentration.

- Nuclear fission is the splitting of a nucleus into daughter nuclides.

- Nuclear power plants generate electricity by using nuclear fission to generate heat that is used to boil water and create steam. The steam turns a turbine on a generator to produce electricity.

- Nuclear fusion is the combination of two light nuclei to form a heavier one.

- In stars, hydrogen atoms undergo fusion to form helium atoms.

- The rem is a weighted measure of radiation exposure that accounts for the ionizing power of different types of radiation.

- Radioactivity is used in medicine for the diagnosis and treatment of disease. Isotope scanning is used to find and identify cancerous tumors. Radiotherapy is used kill cancerous tumors.

Chapter 17

SKILLBUILDER PROBLEMS AND SOLUTIONS

SKILLBUILDER 17.1   Writing Nuclear Equations for Alpha (α) Decay

Write a nuclear equation for the alpha decay of polonium-216.

Solution:
Begin with the symbol for polonium-216 on the left side of the equation and the symbol for an alpha particle on the right side.

$$^{216}_{84}\text{Po} \rightarrow {}^{x}_{y}? + {}^{4}_{2}\text{He}$$

Equalize the sum of the mass numbers and the sum of the atomic numbers on both sides of the equation by writing the appropriate mass number and atomic number for the unknown daughter nuclide.

$$^{216}_{84}\text{Po} \rightarrow {}^{212}_{82}? + {}^{4}_{2}\text{He}$$

Deduce the identity of the unknown daughter nuclide from the atomic number and write its symbol. Since the atomic number is 82, the daughter nuclide must be lead (Pb).

$$^{216}_{84}\text{Po} \rightarrow {}^{212}_{82}\text{Pb} + {}^{4}_{2}\text{He}$$

SKILLBUILDER 17.2   Writing Nuclear Equations for Beta (β) Decay

Write a nuclear equation for the beta decay of actinium-228.

Solution:
Begin with the symbol for actinium-228 on the left side of the equation and the symbol for a beta particle on the right side.

$$^{228}_{89}\text{Ac} \rightarrow {}^{x}_{y}? + {}^{0}_{-1}\text{e}$$

Equalize the sum of the mass numbers and the sum of the atomic numbers on both sides of the equation by writing the appropriate mass number and atomic number for the unknown daughter nuclide.

$$^{228}_{89}\text{Ac} \rightarrow {}^{228}_{90}? + {}^{0}_{-1}\text{e}$$

Deduce the identity of the unknown daughter nuclide from the atomic number and write its symbol. Since the atomic number is 90, the daughter nuclide must be thorium (Th).

Radioactivity and Nuclear Chemistry

$$^{228}_{89}\text{Ac} \rightarrow {}^{228}_{90}\text{Th} + {}^{0}_{-1}\text{e}$$

SKILLBUILDER PLUS

Write three nuclear equations to represent the nuclear decay sequence that begins with the alpha decay of uranium-235 followed by a beta decay of the daughter nuclide and then another alpha decay.

Solution:
Begin with the symbol for uranium-235 on the left side of the equation and the symbol for an alpha particle on the right side.

$$^{235}_{92}\text{U} \rightarrow {}^{x}_{y}? + {}^{4}_{2}\text{He}$$

Equalize the sum of the mass numbers and the sum of the atomic numbers on both sides of the equation by writing the appropriate mass number and atomic number for the unknown daughter nuclide.

$$^{235}_{92}\text{U} \rightarrow {}^{231}_{90}? + {}^{4}_{2}\text{He}$$

Deduce the identity of the unknown daughter nuclide from the atomic number and write its symbol. Since the atomic number is 90, the daughter nuclide must be thorium (Th).

$$^{235}_{92}\text{U} \rightarrow {}^{231}_{90}\text{Th} + {}^{4}_{2}\text{He}$$

We then write the symbol for thorium-231 on the left side of the equation and the symbol for a beta particle on the right side.

$$^{231}_{90}\text{Th} \rightarrow {}^{x}_{y}? + {}^{0}_{-1}\text{e}$$

Equalize the sum of the mass numbers and the sum of the atomic numbers on both sides of the equation by writing the appropriate mass number and atomic number for the unknown daughter nuclide.

$$^{231}_{90}\text{Th} \rightarrow {}^{231}_{91}? + {}^{0}_{-1}\text{e}$$

Deduce the identity of the unknown daughter nuclide from the atomic number and write its symbol. Since the atomic number is 91, the daughter nuclide must be protactinium (Pa).

$$^{231}_{90}\text{Th} \rightarrow {}^{231}_{91}\text{Pa} + {}^{0}_{-1}\text{e}$$

Chapter 17

Finally, write the symbol for protactinium-231 on the left side of the equation and the symbol for an alpha particle on the right side.

$$^{231}_{91}Pa \rightarrow \, ^{x}_{y}? + \, ^{4}_{2}He$$

Equalize the sum of the mass numbers and the sum of the atomic numbers on both sides of the equation by writing the appropriate mass number and atomic number for the unknown daughter nuclide.

$$^{231}_{91}Pa \rightarrow \, ^{227}_{89}? + \, ^{4}_{2}He$$

Deduce the identity of the unknown daughter nuclide from the atomic number and write its symbol. Since the atomic number is 89, the daughter nuclide must be actinium (Ac).

$$^{231}_{91}Pa \rightarrow \, ^{227}_{89}Ac + \, ^{4}_{2}He$$

The decay sequence is

$$^{235}_{92}U \rightarrow \, ^{231}_{90}Th + \, ^{4}_{2}He$$

$$^{231}_{90}Th \rightarrow \, ^{231}_{91}Pa + \, ^{0}_{-1}e$$

$$^{231}_{91}Pa \rightarrow \, ^{227}_{89}Ac + \, ^{4}_{2}He$$

| SKILLBUILDER 17.3 | Writing Nuclear Equations for Positron Emission |

Write a nuclear equation for the positron emission of sodium-22.

Solution:
Begin with the symbol for sodium-22 on the left side of the equation and the symbol for a positron on the right side.

$$^{22}_{11}Na \rightarrow \, ^{x}_{y}? + \, ^{0}_{+1}e$$

Equalize the sum of the mass numbers and the sum of the atomic numbers on both sides of the equation by writing the appropriate mass number and atomic number for the unknown daughter nuclide.

$$^{22}_{11}Na \rightarrow \, ^{22}_{10}? + \, ^{0}_{+1}e$$

Deduce the identity of the unknown daughter nuclide from the atomic number and write its symbol. Since the atomic number is 10, the daughter nuclide must be neon (Ne).

Radioactivity and Nuclear Chemistry

$$^{22}_{11}\text{Na} \rightarrow {}^{22}_{10}\text{Ne} + {}^{0}_{+1}\text{e}$$

SKILLBUILDER 17.4   Half-life

A radium-226 sample initially contains 0.112 mol. How much radium-226 is left in the sample after 6400 years? The half-life of radium-226 is 1600 years.

Solution:
It is easiest to draw a table showing the amount of radium-226 as a function of the number of half-lives. For each half-life, simply divide the amount of radium-226 by two.

Number of Half-lives	Time in Years	Amount of Radium-226
0	0	1.12×10^{-1} mol
1	1600	5.60×10^{-2} mol
2	3200	2.80×10^{-2} mol
3	4800	1.40×10^{-2} mol
4	6400	7.00×10^{-3} mol

Therefore, there will be 7.00×10^{-3} mol radium-226 left after 1600 years.

SKILLBUILDER 17.5   Radiocarbon Dating

An ancient scroll is claimed to have originated from Greek scholars in about 500 B.C. A measure of its carbon-14 content reveals it to contain 100.0% of that found in living organisms. Is the scroll authentic?

Solution:
Since the scroll contains as much carbon-14 as a living organism, it is from the same time period as a living organism. The scroll is not authentic.

SELF-TEST QUESTIONS

A. Match the following terms with the phrases below.

- alpha (α) radiation
- beta (β) radiation
- chain reaction
- critical mass
- daughter nuclide
- film-badge dosimeter
- Geiger-Müller counter
- half-life
- ionizing power
- isotope scanning
- nuclear equation
- penetrating power
- positron emission
- radioactive
- radiocarbon dating
- scintillation counter

Chapter 17

1. Characteristic of atoms that emit particles from their nuclei
2. Radiation that occurs when an unstable nucleus emits a small piece of itself composed of 2 protons and 2 neutrons
3. Equation that represents a nuclear process
4. Atoms formed from a parent nuclide
5. Ability of radiation to ionize other molecules and atoms
6. Ability to penetrate matter
7. Electron emitted from a nucleus
8. Emission that occurs when an unstable nucleus emits a positron
9. Radiation detector consisting of a photographic film held in a small case
10. Radioactivity detector in which radioactive particles pass through an argon-filled chamber
11. Radiation detector in which radioactive particles pass through a material that emits ultraviolet or visible light in response to excitation by radioactive particles
12. Time required for one-half of the parent nuclides in a radioactive sample to decay to the daughter nuclides
13. Dating technique that measures carbon-14 content
14. Self-amplifying reaction
15. Mass of a nuclide necessary to produce a self-sustaining reaction
16. Technique used in medicine in which a radioactive isotope is introduced into the body and the emitted radiation detected

B. True/False

1. Antoine-Henri Becquerel was the first to discover radioactivity.
2. Marie Curie received the Nobel Prize in chemistry in 1911 for her discovery of polonium and radium.
3. An alpha particle has a mass number of 4.
4. Alpha particles have a high penetrating power.
5. A beta particle has a mass number of –1.
6. Gamma radiation has no mass.
7. A positron has a charge of –1.
8. After two half-lives the entire amount of a radioactive sample will have decayed.
9. Radiocarbon dating of an artifact is based on its concentration of carbon-12.
10. Nuclear fusion involves splitting of an atom into smaller nuclides.

C. Multiple Choice

1. Write the symbol for the isotope of cobalt that has 33 neutrons.
    a) $^{59}_{27}Co$
    b) $^{60}_{27}Co$
    c) $^{33}_{27}Co$
    d) $^{59}_{33}Co$

Radioactivity and Nuclear Chemistry

2. How many protons and how many neutrons are in $^{99}_{42}Mo$?
   a) 44 protons, 99 neutrons
   b) 99 protons, 42 neutrons
   c) 42 protons, 57 neutrons
   d) 57 protons, 42 neutrons

3. Which of the following is the symbol for a gamma ray?
   a) $^{0}_{-1}e$
   b) $^{0}_{0}\gamma$
   c) $^{0}_{+1}e$
   d) $^{1}_{1}p$

4. Which of the following is the symbol for a positron?
   a) $^{1}_{1}p$
   b) $^{0}_{+1}e$
   c) $^{0}_{-1}e$
   d) $^{1}_{0}n$

5. Which of the following is the daughter nuclide produced when polonium-210 emits an alpha particle?
   a) $^{214}_{82}Pb$
   b) $^{206}_{82}Pb$
   c) $^{214}_{86}Rn$
   d) $^{206}_{86}Rn$

6. Which of the following is the daughter nuclide produced when carbon-14 emits a beta particle?
   a) $^{14}_{5}B$
   b) $^{13}_{6}C$
   c) $^{13}_{7}N$
   d) $^{14}_{7}N$

7. Which of the following is the daughter nuclide produced when potassium-38 emits a positron?
   a) $^{38}_{18}Ar$
   b) $^{39}_{19}K$
   c) $^{38}_{20}Ca$
   d) $^{39}_{20}Ca$

Chapter 17

8. Strontium-90 decays by beta emission with a half-life of 28 years. How much of a 10.0-g sample of strontium-90 remains after 84 years?
   a) 5.00 g
   b) 2.50 g
   c) 1.25 g
   d) 0.00 g

9. The half-life of oxygen-15 is 2 minutes. How long will it take for 0.800 mol of oxygen-15 to decay to 0.100 mol?
   a) 2 minutes
   b) 4 minutes
   c) 6 minutes
   d) 8 minutes

10. The half-life of carbon-14 is 5730 years. An artifact has a carbon-14 concentration that is 1/16 the concentration of carbon-14 in a living organism. How old is the artifact?
    a) 5730 years
    b) 11,460 years
    c) 17,190 years
    d) 22,920 years

Radioactivity and Nuclear Chemistry

D. Crossword Puzzle

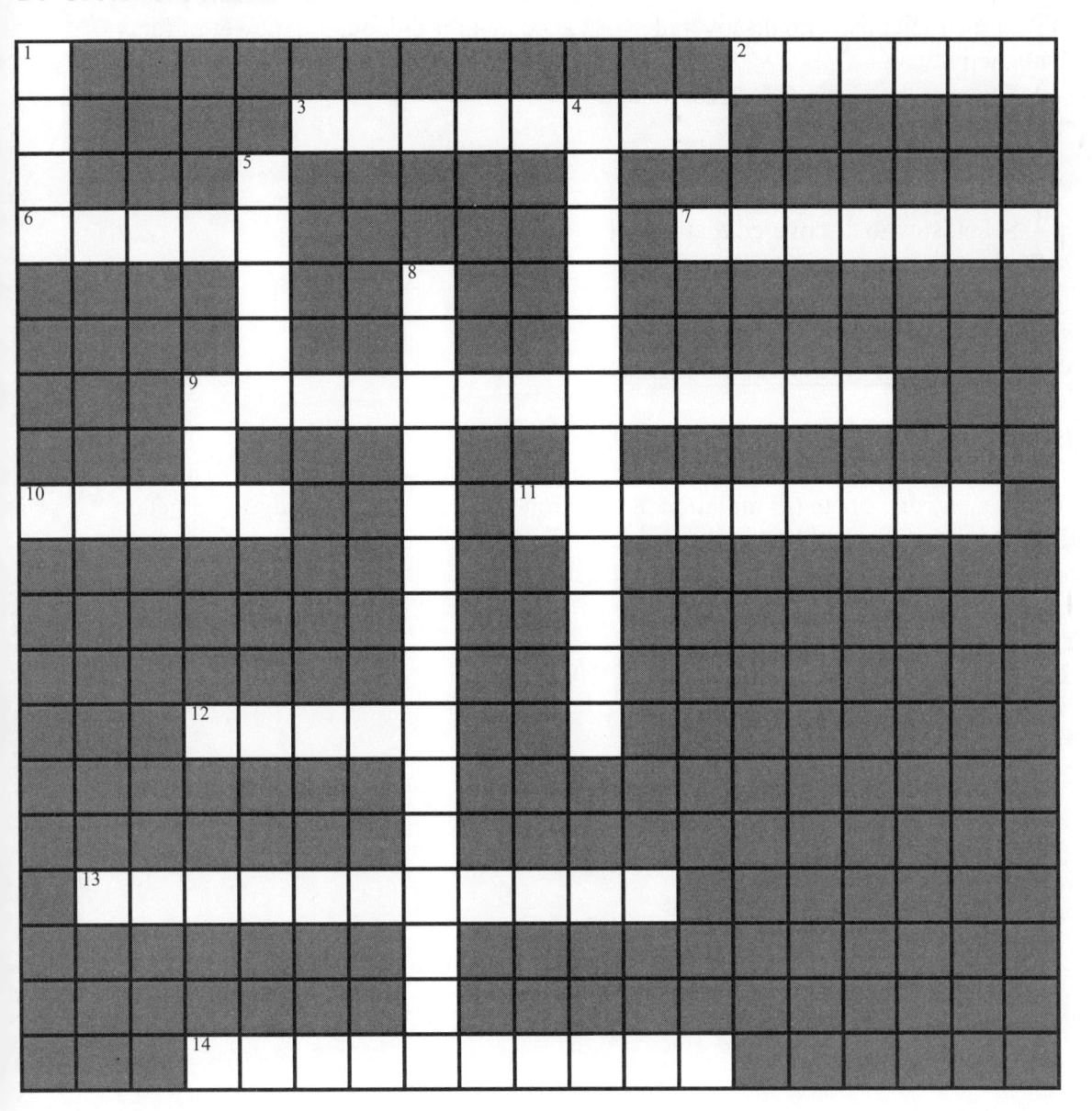

ACROSS

2. Combination of two light nuclei to form a heavier one
3. Positively charged particle emitted from a proton as it converts into a neutron
6. Particle identical to a helium nucleus
7. Splitting of an atom
9. Emission of tiny, invisible particles by the nuclei of certain atoms
10. Italian physicist who bombarded uranium with neutrons and detected beta

DOWN

1. Radiation that occurs when an unstable nucleus emits an electron
4. Medical technique in which gamma rays are focused on internal tumors to kill them
5. Radiation consisting of high energy photons
8. Long-lived emission of light that sometimes follows the absorption of light by some atoms and molecules

Chapter 17

emission
11. French scientist who discovered radioactivity
12. Scientist who discovered radium and polonium
13. Physicist who led the Manhattan Project
14. Scientist who discovered that nuclei of radioactive atoms emit particles

9. Measure of radiation exposure that accounts for the ionizing power of the different types of radiation

ANSWERS TO SELF-TEST QUESTIONS

A. Matching
1. radioactive 2. alpha (α) radiation 3. nuclear equation 4. daughter nuclide 5. ionizing power
6. penetrating power 7. beta (β) radiation 8. positron emission 9. film-badge dosimeter
10. Geiger-Müller counter 11. scintillation counter 12. half-life 13. radiocarbon dating
14. chain reaction 15. critical mass 16. isotope scanning

B. True/False
1. T 2. T 3. T 4. F 5. F 6. T 7. F 8. F 9. F 10. F

C. Multiple Choice
1. b 2. c 3. b 4. b 5. b 6. d 7. a 8. c 9. c 10. d

D. Crossword Puzzle

	1	2	3	4	5	6	7	8	9

Across and Down answers filled in the grid:

- 1. BETA (down)
- 2. FUSION (across)
- 3. POSITRON (across)
- 4. RADIATION (down)
- 5. GAMMA (down)
- 6. ALPHA (across)
- 7. FISSION (across)
- 8. PHOSPHORESCENCE (down)
- 9. RADIOACTIVITY (across)
- 10. FERMI (across)
- 11. BECQUEREL (across)
- 12. CURIE (across)
- 13. OPPENHEIMER (across)
- 14. RUTHERFORD (across)

261

Chapter 17

Organic Chemistry 18

CHAPTER OVERVIEW

Chapter 18 introduces organic chemistry. Names, structures, and properties of alkanes, alkenes, alkynes, and aromatic hydrocarbons are examined. Reactions of hydrocarbons are presented. Families of compounds and their functional groups are discussed. Polymers and their commercial importance are described.

CHAPTER OBJECTIVES

After reading and studying the text, students should be able to:

1. Explain the difference between organic and inorganic compounds as it was viewed at the end of the eighteenth century.
2. Explain the concept of vitalism.
3. Differentiate between saturated hydrocarbons and unsaturated hydrocarbons.
4. List the main uses for hydrocarbons.
5. Explain the difference between a molecular formula, a structural formula, and a condensed structural formula.
6. Explain the difference between *n*-alkanes and branched alkanes.
7. Write structural formulas for hydrocarbon isomers.
8. Name alkanes from structural formulas.
9. Differentiate between alkanes, alkenes, and alkynes based on their molecular formulas.
10. Describe the differences in bonding between alkanes, alkenes, and alkynes.
11. Name alkenes.
12. Draw structural formulas of alkenes.
13. Name alkynes.
14. Draw structural formulas of alkynes.
15. Name aromatic compounds.
16. Draw structural formulas of aromatic compounds.
17. Write general formulas for alcohols, ethers, aldehydes, ketones, carboxylic acids, esters, and amines.
18. Identify the functional groups in a structural formula.
19. Name alcohols.
20. Draw structures of alcohols.
21. Give examples of common alcohols.
22. Name ethers.
23. Draw structures of ethers.
24. Give an example of a common ether and its uses.

Chapter 18

25. Name aldehydes and ketones.
26. Draw structures of aldehydes and ketones.
27. Give examples of common aldehydes and ketones.
28. Name carboxylic acids.
29. Draw structures of carboxylic acids.
30. Give examples of common carboxylic acids.
31. Name esters.
32. Draw structures of esters.
33. Give examples of common esters.
34. Name amines.
35. Draw structures of amines.
36. Give an example of a common amine.
37. Explain what a polymer is.
38. State the difference between a polymer and a copolymer.
39. Explain the difference between a condensation polymer and an addition polymer.
40. Draw structures of polymers given the structures of the monomers from which they form.

CHAPTER IN REVIEW

- Organic chemistry is the study of carbon-containing compounds and their reactions.

- Hydrocarbons are molecules that contain only carbon and hydrogen.

- Hydrocarbons in which all carbon-carbon bonds are single bonds are called alkanes. The general formula for an alkane is C_nH_{2n+2}.

- Alkanes are saturated hydrocarbons; they are saturated with hydrogen.

- Alkanes in which the carbon atoms are bonded in a straight chain are called normal alkanes, or *n*-alkanes.

- Methane, ethane, propane, butane, pentane, hexane, heptane, and octane are normal alkanes with one, two, three, four, five, six, seven, and eight carbon atoms, respectively.

- The boiling points of normal alkanes increase with an increase in the number of carbon atoms.

- Branched alkanes are alkanes in which the carbon atoms form a branched structure.

- Isomers are molecules with the same molecular formula but different structures.

- In the IUPAC (International Union of Pure and Applied Chemistry) system, the longest continuous chain of carbon atoms determines the base name of the compound. Groups of carbon atoms branching off the base chain are called alkyl groups. A substituent is an atom or group of atoms that has been substituted for a hydrogen atom.

- To name an alkane, count the number of carbon atoms in the longest continuous carbon chain to determine the base name of the compound. Name each substituent. Beginning with the end closest to the branching, number the base chain and assign a number to each substituent. If two substituents occur at equal distances from each end, go to the next substituent to determine from which end to start numbering. Write the name of the compound in the format (substituent number) – (substituent name)(base name). If there are two or more substituents, give each one a number and list them alphabetically with hyphens between words and numbers. If a compound has two or more identical substituents, use the prefixes *di-*, *tri-*, or *tetra-* before their names. Separate the numbers indicating their positions from each other with a comma. The prefixes are ignored when listing alphabetically.

- Alkenes are hydrocarbons that contain a carbon-carbon double bond. The general formula for an alkene is C_nH_{2n}.

- Alkynes are hydrocarbons that contain a carbon-carbon triple bond. The general formula for an alkyne is C_nH_{2n-2}.

- Alkenes and alkynes are unsaturated hydrocarbons—they are not loaded to capacity with hydrogen.

- Alkanes and alkynes are named in the same way as alkanes except that the base chain is the longest continuous chain that contains the double or triple bond. The base name has the ending *–ene* for alkenes and *–yne* for alkynes. The base chain is numbered to give the double or triple bond the lowest possible number. A number indicating the position of the double or triple bond is inserted just before the base name.

- All hydrocarbons undergo combustion reactions.

- Alkanes undergo substitution reactions.

- Alkenes and alkynes undergo addition reactions.

- Compounds that contain benzene rings are called aromatic compounds.

- Monosubstituted benzenes are benzene rings in which one of the hydrogen atoms has been substituted. If the substituent is very large, the compound is named by treating the benzene ring as a substituent called a phenyl group.

- Disubstituted benzenes are benzene rings in which two hydrogen atoms have been substituted. The substituents are numbered and listed alphabetically. The order of numbering is determined by the alphabetical order of the substituents.

Chapter 18

- A functional group is a characteristic atom or group of atoms that has been inserted in to a hydrocarbon. The functional group alters the properties of the compound significantly. A group of organic compounds with the same functional group forms a family.

- Common families of functional groups are alcohols, ethers, aldehydes, ketones, carboxylic acids, esters, and amines.

- Alcohols have the general formula R–OH. Alcohols are named in the same way as alkanes except that the base chain is the longest continuous carbon chain that contains the –OH functional group. The base name has the ending –ol. The base chain is numbered to give the –OH group the lowest possible number. A number indicating the position of the –OH group is inserted just before the base name.

- Ethers are organic compounds in which two alkyl groups are bonded to an oxygen atom and have the general formula R–O–R.

- Aldehydes, ketones, carboxylic acids, and esters contain a carbonyl group.

- The general formula for an aldehyde is RCHO. Aldehydes are named according to the number of carbon atoms in the longest continuous carbon chain that contains the carbonyl group. The base name is formed by dropping the –e from the corresponding alkane and adding the ending –al.

- The general formula for a ketone is RCOR. Ketones are named according to the number of carbon atoms in the longest continuous carbon chain that contains the carbonyl group. The base name is formed by dropping the –e from the corresponding alkane and adding the ending –one. The chain is numbered to give the carbonyl group the lowest possible number.

- The general formula for a carboxylic acid is RCOOH. Carboxylic acids are named according to the number of carbon atoms in the longest continuous carbon chain containing the –COOH functional group. The base name is formed by dropping the –e from the corresponding alkane and adding the ending –oic acid.

- The general formula for an ester is RCOOR. Esters are named as if they were derived from a carboxylic acid by replacing the H of the –COOH group with an alkyl group. The R group from the parent acid forms the base name of the compound. The –ic ending on the name of the corresponding carboxylic acid is changed to –ate. The R group that replaced the H of the carboxylic acid is named as an alkyl group.

- Amines are organic compounds containing nitrogen. Amines are derivatives of ammonia, NH_3, in which one or more of the hydrogen atoms are replaced by alkyl groups. Amines are named according to the alkyl groups attached to the nitrogen and given the ending –amine.

Organic Chemistry

- Polymers are long chain-like molecules composed of repeating units. The individual repeating units are called monomers. Copolymers consist of two different monomers.

- Condensation polymers are polymers that eliminate an atom or group of atoms during polymerization.

- Dimers are the product of the reaction of two monomers.

SKILLBUILDER PROBLEMS AND SOLUTIONS

SKILLBUILDER 18.1 **Differentiating Between Alkanes, Alkenes, and Alkynes Based on Their Molecular Formulas**

Based on the molecular formula, determine whether the following noncyclical hydrocarbons are alkanes, alkenes, or alkynes.

a) C_6H_{12}
b) C_8H_{14}
c) C_5H_{12}

Solution:
a) C_6H_{12}
The number of carbons is 6; therefore, $n = 6$. If $n = 6$, then 12 is $2n$. The molecule must be an alkene.

b) C_8H_{14}
The number of carbons is 8; therefore, $n = 8$. If $n = 8$, then 14 is $2n - 2$. The molecule must be an alkyne.

c) C_5H_{12}
The number of carbons is 5; therefore, $n = 5$. If $n = 5$, then 12 is $2n + 2$. The molecule must be an alkane.

SKILLBUILDER 18.2 **Writing Formulas for *n*-Alkanes**

Write the structural and condensed structural formula for C_5H_{12}.

Solution:
Structural formula:
The first step is to write out the carbon backbone with 5 carbons in it.

$$C - C - C - C - C$$

Chapter 18

The next step is to add hydrogen atoms so that all carbon atoms have 4 bonds.

```
    H   H   H   H   H
    |   |   |   |   |
H — C — C — C — C — C — H
    |   |   |   |   |
    H   H   H   H   H
```

Condensed structural formula:
To write the condensed structural formula, write the hydrogen atoms bonded to each carbon directly to the right of the carbon atom. Use subscripts to indicate the correct number of hydrogen atoms.

$CH_3CH_2CH_2CH_2CH_3$

SKILLBUILDER 18.3   **Writing Structural Formulas for Isomers**

Draw the three isomers for pentane.

Solution:
To start, always draw the carbon backbone. The first isomer is the straight-chain isomer.

C — C — C — C — C

Then determine the carbon backbone structure of the other isomers by arranging the carbon atoms in two other unique ways.

```
C — C — C — C — C                    C
                                     |
                                 C — C — C
                                     |
    C — C — C — C                    C
        |
        C
```

Finally, fill in all the hydrogen atoms so that each carbon atom has four bonds.

Organic Chemistry

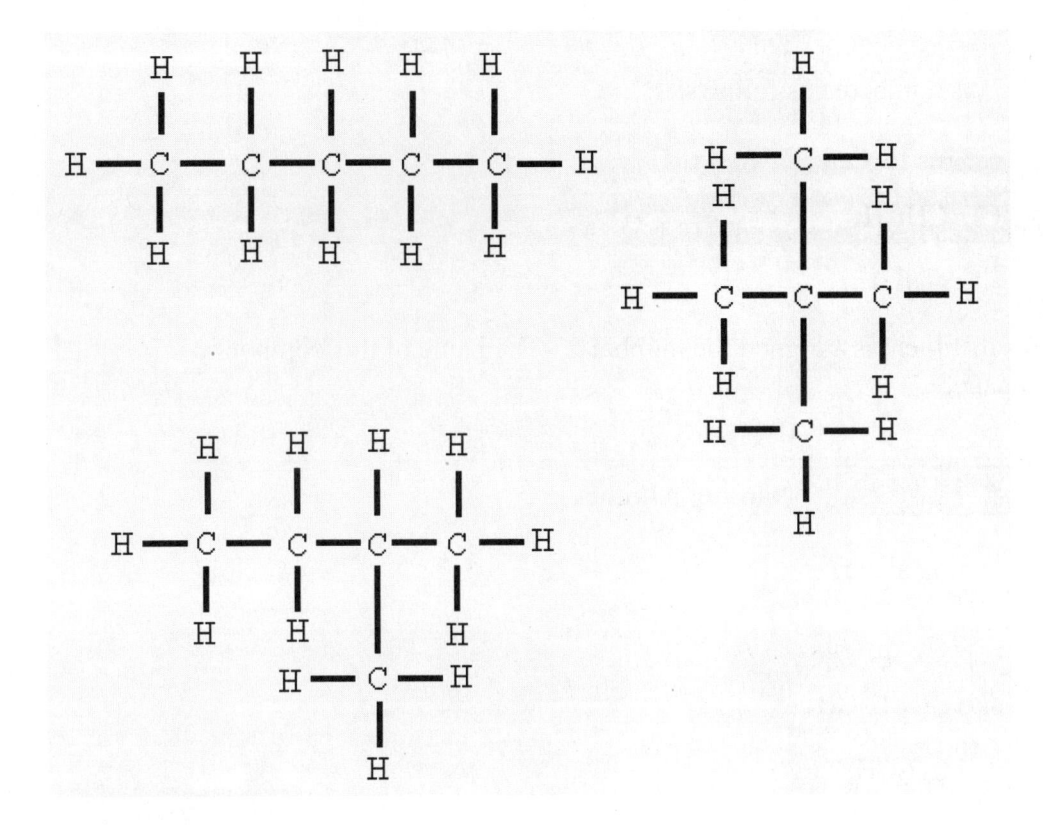

H H H H H
| | | | |
H—C—C—C—C—C—H
| | | | |
H H H H H

            H
            |
       H—C—H
        H | H
        | |
    H—C—C—C—H
        | |
        H | H
       H—C—H
           |
           H

H H H H
| | | |
H—C—C—C—C—H
| | | |
H H | H
     H—C—H
        |
        H

SKILLBUILDER 18.4   Naming Alkanes

Name the following alkane.

    CH$_3$CHCH$_3$
         |
        CH$_3$

Solution:
This compound has 3 carbon atoms in its longest continuous chain.

    CH$_3$CHCH$_3$
         |
        CH$_3$

The correct prefix is *prop-*. The base name is *propane*.

This compound has one substituent named *methyl*.

    CH$_3$CHCH$_3$
         |
        CH$_3$

Chapter 18

The base chain is numbered as follows:

$$\begin{array}{c} \overset{1}{C}H_3\overset{2}{C}H\overset{3}{C}H_3 \\ | \\ CH_3 \end{array}$$

The methyl substituent is assigned the number 2. The name of this compound is 2-methylpropane.

SKILLBUILDER 18.5   Naming Alkanes

Name the following alkane.

$$\begin{array}{c} CH_3 \\ | \\ CH_3CHCHCH_2CH_3 \\ | \\ CH_2CH_3 \end{array}$$

Solution:
This compound has 5 carbon atoms in its longest continuous chain.

$$\begin{array}{c} CH_3 \\ | \\ \mathbf{CH_3CHCHCH_2CH_3} \\ | \\ CH_2CH_3 \end{array}$$

The correct prefix is *pent-*. The base name is *pentane*.

This compound has one substituent named *methyl* and one named *ethyl*.

$$\begin{array}{c} \mathbf{CH_3} \text{ methyl} \\ | \\ CH_3CHCHCH_2CH_3 \\ | \\ \mathbf{CH_2CH_3} \text{ ethyl} \end{array}$$

The base chain is numbered as follows:

$$\begin{array}{c} CH_3 \\ \overset{1}{C}H_3\overset{|}{\underset{2}{C}H}\overset{3}{C}H\overset{4}{C}H_2\overset{5}{C}H_3 \\ | \\ CH_2CH_3 \end{array}$$

Organic Chemistry

The methyl substituent is assigned the number 2 and the ethyl substituent is assigned the number 3. Ethyl is listed before methyl because substituents are listed in alphabetical order. The name of this compound is 3-ethyl-2-methylpentane.

SKILLBUILDER 18.6 Naming Alkanes

Name the following alkane.

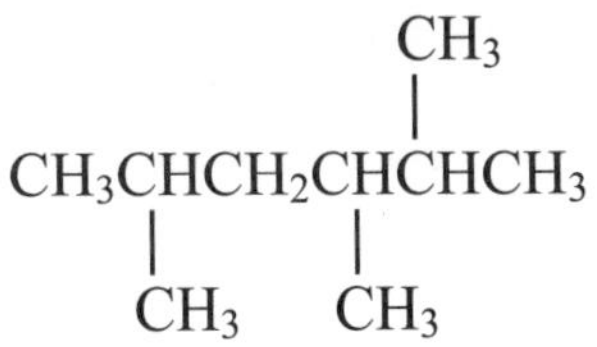

$$\text{CH}_3\text{CHCH}_2\text{CHCHCH}_3$$
with CH$_3$ branches on carbons 2, 3, and the middle carbon (with CH$_3$ above).

Solution:
The longest continuous carbon chain has 6 carbons. Therefore, the base name is *hexane*.

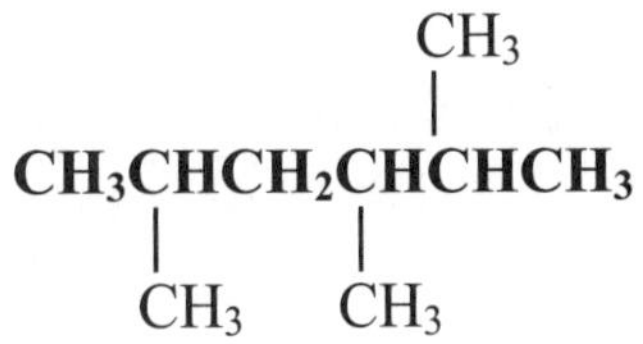

This compound has three substituents, all of which are named *methyl*.

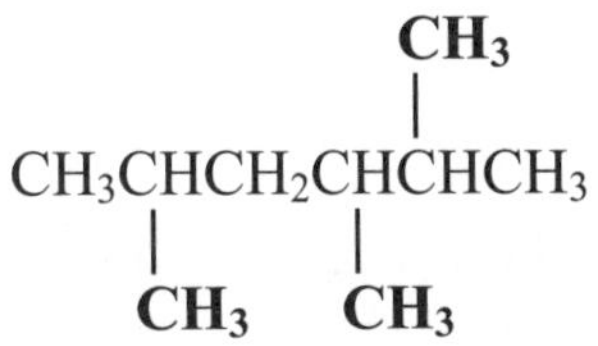

Two substituents are equidistant from the ends, so we look at the next substituent to determine from which end to start numbering.

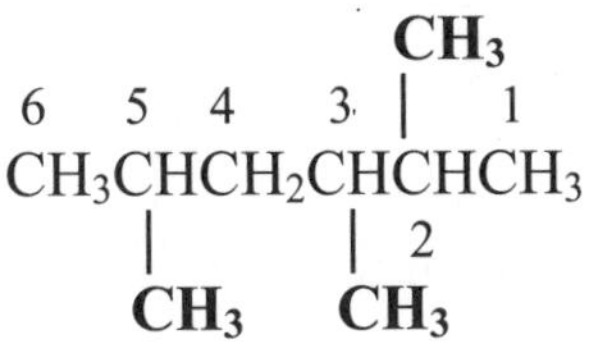

Numbering: 6 5 4 3 2 1 from left to right, with methyl groups at positions 2, 3, and 5.

Since this compound has three identical substituents, we use the prefix *tri-* before their names. The name of this compound is 2,3,5-trimethylhexane.

271

Chapter 18

| SKILLBUILDER 18.7 | **Naming Alkenes and Alkynes** |

Name the following alkene and alkyne.

```
          H₃C
           |
H₃C—C≡C—C—CH₃
           |
          H₃C
```

```
                    CH₃
                     |
                    H₂C
                     |
H₃C—CH—CH₂—CH—CH—CH=CH₂
     |        |
    CH₃      H₃C
```

Solution:
In the first compound, the longest continuous chain containing the triple bond has 5 carbons. The base name is therefore *pentyne*.

```
          H₃C
           |
H₃C—C≡C—C—CH₃
           |
          H₃C
```

The two substituents are both *methyl*.

We number the chain so that the triple bond has the lowest number. The triple bond is assigned the number 2. Both methyl groups are assigned the number 4.

```
              H₃C
       1  2  3 |4  5
      H₃C—C≡C—C—CH₃
               |
              H₃C
```

Both substituents are identical, so we use the prefix *di*-. The name of this compound is 4,4-dimethyl-2-pentyne.

In the second compound, the longest continuous chain containing the double bond has 7 carbons. The base name is therefore *heptene*.

Organic Chemistry

```
              CH3
              |
              H2C
              |
H3C—CH—CH2—CH—CH—CH=CH2
     |         |
     CH3      H3C
```

This compound has two *methyl* substituents and one *ethyl* substitutent.

```
              CH3
              |
              H2C
              |
H3C—CH—CH2—CH—CH—CH=CH2
     |         |
     CH3      H3C
```

We number the chain so that the double bond has the lowest number. The double bond is assigned the number 1. The ethyl group is assigned the number 3. The methyl groups are assigned the numbers 4 and 6.

```
                  CH3
                  |
                  H2C
 7   6    5   4   |3   2    1
H3C—CH—CH2—CH—CH—CH=CH2
     |         |
     CH3      H3C
```

Ethyl is listed before methyl because substituents are listed in alphabetical order. Two of the substituents are identical, so we use the prefix *di*-. The name of this compound is 3-ethyl-4,6-dimethyl-1-heptene.

SKILLBUILDER 18.8   Naming Aromatic Compounds

Name the following compound.

Chapter 18

Solution:
Benzene derivatives are named using the general form (name of substituent)*benzene*. This compound has two substituents named *bromo*. Because this derivative has two substituents, the substituents are numbered. One bromo substituent is assigned the number 1; the other bromo substituent is assigned the number 3. Or, in place of numbering, we use the prefix *meta-* or *m-*. Since both substituents are identical, we use the prefix *di-*. The name of this compound is 1,3-dibromobenzene or *meta*-dibromobenzene or *m*-dibromobenzene.

SELF-TEST QUESTIONS

A. **Match the following terms with the phrases below.**

addition polymer
addition reaction
alkane
alkyl group
aromatic rings
base chain
branched alkane
carbonyl group
carboxylic acid
condensation polymer
condensed structural formula
disubstituted benzene

family (of organic compounds)
fossil fuel
functional group
monosubstituted benzene
normal alkane (*n*-alkane)
organic chemistry
organic molecule
phenyl group
structural formula
substitution reaction
unsaturated hydrocarbon
vital force

1. Molecule that contains carbon
2. Study of carbon-containing compounds and their reactions
3. Mystic or supernatural power that allows living organisms to produce organic compounds
4. Hydrocarbon fuel
5. Hydrocarbon containing only single bonds
6. Formula that shows the number and type of each atom in the molecule and the structure of the molecule
7. Shorthand way of writing a structural formula
8. Alkane composed of carbon atoms bonded in a straight chain without any branching
9. Alkane composed of carbon atoms with branching
10. Longest continuous chain of carbon atoms
11. Hydrocarbon group branching off the base chain of an organic compound
12. Hydrocarbon that is not loaded to capacity with hydrogen
13. Reaction in which one or more hydrogen atoms are replaced by one or more other atoms
14. Reaction in which atoms add across a multiple bond
15. Benzene rings
16. Benzene ring in which only of the hydrogen atoms has been substituted
17. Benzene ring as a substituent
18. Benzene ring in which two hydrogen atoms have been substituted
19. Characteristic atom or group of atoms

20. Group of organic compounds with the same functional group
21. Functional group in which a carbon atom is double bonded to an oxygen atom and singly bonded to two other groups
22. Organic compound with the general formula RCOOH
23. Polymer in which the monomers simply link together without the elimination of any atoms
24. Polymer that eliminates an atom or small group of atoms during polymerization

B. True/False

1. Carbon atoms can bond to one another to form chain, branched, and ring structures.
2. Butane has the formula C_4H_8.
3. Alkanes undergo addition reactions.
4. Alkenes and alkynes are saturated hydrocarbons.
5. Benzene is an aromatic compound.
6. Isopropyl alcohol is highly toxic.
7. Aldehydes have the general formula R–OH.
8. Ketones contain a carbonyl group.
9. The active ingredient in vinegar is acetic acid (ethanoic acid).
10. Polyethylene is an addition polymer.

C. Multiple Choice

1. Which of the following noncyclical hydrocarbons is an alkane?
   a) C_5H_{10}
   b) C_6H_{14}
   c) C_7H_{12}
   d) C_8H_{16}

2. Which of the following alkanes has the highest boiling point?
   a) Methane
   b) Ethane
   c) Propane
   d) Butane

3. Which of the following is the formula for heptane?
   a) C_5H_{12}
   b) C_6H_{14}
   c) C_7H_{16}
   d) C_8H_{18}

4. Which of the following is the name for the alkyl group with the condensed structural formula $-CH_2CH_2CH_2CH_3$?
   a) Methyl
   b) Ethyl
   c) Propyl
   d) Butyl

Chapter 18

5. Which of the following is the condensed structural formula for 3-hexene?
   a) CH₃CH₂CH₂CH₂CH₂CH₃
   b) CH₂CHCH₂CH₂CH₂CH₃
   c) CH₃CHCHCH₂CH₂CH₃
   d) CH₃CH₂CHCHCH₂CH₃

6. Which of the following substituted benzenes has a methyl substituent?
   a) Toluene
   b) Phenol
   c) Aniline
   d) Styrene

7. A compound of general formula R–O–R is an
   a) Alcohol
   b) Ether
   c) Aldehyde
   d) Ester

8. What is the condensed general formula for a carboxylic acid?
   a) ROH
   b) RCHO
   c) RCOOH
   d) RCOOR

9. Which class of organic compounds contains nitrogen?
   a) Amines
   b) Ketones
   c) Aldehydes
   d) Alcohols

10. Which of the following is a condensation polymer used in water-protective coatings?
    a) Polyethylene
    b) Polypropylene
    c) Polystyrene
    d) Polyurethane

Organic Chemistry

D. Crossword Puzzle

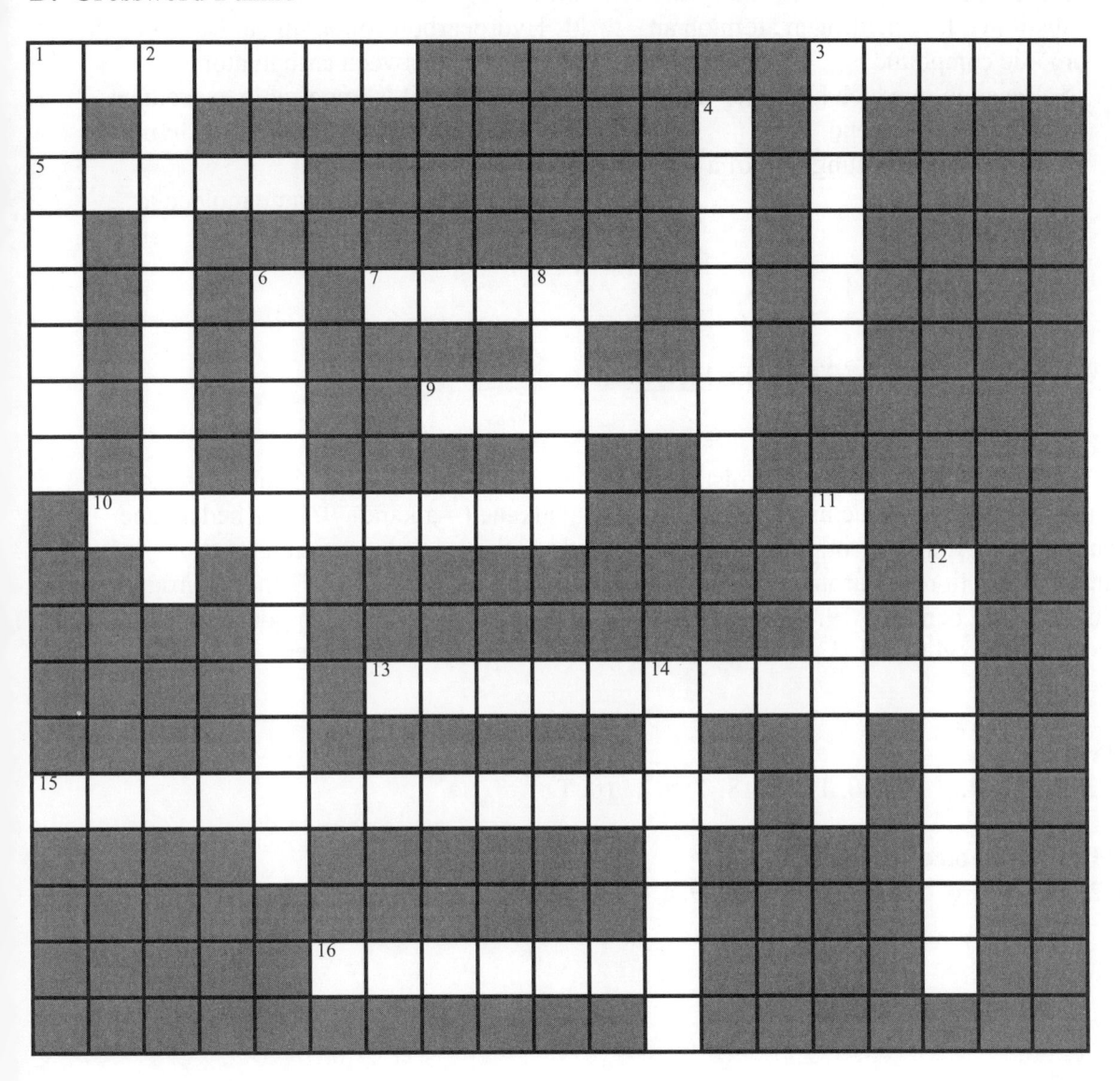

ACROSS

1. Organic compound with an –OH functional group
3. Organic compound containing nitrogen
5. Product that forms from the reaction of two monomers
7. Organic compound with the general formula R–O–R
9. Organic compound with the general formula RCOR
10. Polymer that consists of two different kinds of monomers

DOWN

1. Organic compound with the general formula RCHO
2. Burning of hydrocarbons in the presence of oxygen
3. Hydrocarbon containing at least one triple bond between carbon atoms
4. Long chain-like molecule composed of repeating units
6. Compound that contains only carbon and hydrogen
8. Organic compound with the general

Chapter 18

13. Atom or group of atoms that has been substituted for a hydrogen atom on an organic compound
15. Reaction in which hydrogen adds across a multiple bond
16. Individual repeating unit of a polymer

formula RCOOR
11. Hydrocarbon containing at least one double bond between carbon atoms
12. Belief that it is impossible to produce an organic compound outside of a living organism
14. Molecules with the same molecular formula but different structures

ANSWERS TO SELF-TEST QUESTIONS

A. Matching
1. organic molecule 2. organic chemistry 3. vital force 4. fossil fuel 5. alkane 6. structural formula 7. condensed structural formula 8. normal alkane (*n*-alkane) 9. branched alkane 10. base chain 11. alkyl group 12. unsaturated hydrocarbon 13. substitution reaction 14. addition reaction 15. aromatic rings 16. monosubstituted benzene 17. phenyl group 18. disubstituted benzene 19. functional group 20. family (of organic compounds) 21. carbonyl group 22. carboxylic acid 23. addition polymer 24. condensation polymer

B. True/False
1. T 2. F 3. F 4. F 5. T 6. T 7. F 8. T 9. T 10. T

C. Multiple Choice
1. b 2. d 3. c 4. d 5. d 6. a 7. b 8. c 9. a 10. d

Organic Chemistry

D. Crossword Puzzle

Across:
1. ALCOHOL
3. AMINE
5. DIMER
7. ETHER
9. KETONE
10. COPOLYMER
13. SUBSTITUENT
15. HYDROGENATION
16. MONOMERS

Down:
1. ALDEHYDE
2. COBUSTIR (COBUSTIRN)
3. ALKYNE
4. POLYMER
6. HYDROCARBON
8. ESTER
11. ALKYL
12. VITALISM
14. ISOMERS

Chapter 18

Biochemistry 19

CHAPTER OVERVIEW

The subject of Chapter 19 is biochemistry. The structure and function of the main chemical components of the living cell—carbohydrates, lipids, proteins, and nucleic acids—are examined. The process of DNA replication is described.

CHAPTER OBJECTIVES

After reading and studying the text, students should be able to:

1. State the main goals of the human genome project.
2. Give two results of the human genome project.
3. List the main chemical components of the cell.
4. Write the general formula for a carbohydrate.
5. List the functions of carbohydrates in living organisms.
6. State the differences between simple carbohydrates and complex carbohydrates.
7. Describe the differences between starch and cellulose and glycogen.
8. Classify carbohydrates as monsaccharides, disaccharides, or trisaccharides.
9. Classify carbohydrates as trioses, tetraoses, or pentoses.
10. Describe the main functions of lipids in living organisms.
11. Draw the general structure for a fatty acid.
12. Explain the difference between a saturated fatty acid and an unsaturated fatty acid.
13. Draw the general structure of a triglyceride.
14. Identify triglycerides.
15. Classify triglycerides as saturated or unsaturated.
16. Describe the main functions of phospholipids and glycolipids in living organisms.
17. List common steroids.
18. List the main functions of proteins in living organisms.
19. Draw the general structure for an amino acid.
20. Show how amino acids react to form peptide bonds.
21. Draw structures for tripeptides.
22. Describe the α-helix and β-pleated sheet structures.
23. Classify patterns in protein structure as primary, secondary, tertiary, or quaternary.
24. Describe the types of interactions that maintain primary, secondary, tertiary, and quaternary protein structure.
25. Describe the main function of nucleic acids in living organisms.
26. State the two main types of nucleic acids.
27. Explain what a codon is.

Chapter 19

28. Explain what a gene is.
29. State the main function of chromosomes.
30. State the number of chromosomes in humans.
31. List the four different bases that occur within DNA.
32. Identify nucleotides and their bases.
33. Give the complementary base of adenine, thymine, cytosine, and guanine.
34. Draw complementary strands of DNA strands.
35. Describe the process by which DNA replicates.
36. Explain how protein synthesis occurs.

CHAPTER IN REVIEW

- The cell is the smallest structural unit of a living organism that has the properties associated with life.

- The nucleus of the cell contains the genetic material.

- The region between the nucleus and the cell membrane is called the cytoplasm. The cytoplasm contains a number of specialized structures that carry out much of the cell's functions.

- The main chemical components of the cell are carbohydrates, lipids, proteins, and nucleic acids.

- Carbohydrates are responsible for short-term energy storage in living organisms. They often have the general formula $(CH_2O)_n$. Carbohydrates are aldehydes or ketones with multiple –OH groups.

- Monosaccharides are carbohydrates that cannot be broken down into simpler carbohydrates. The general names for monosaccharides have a prefix that corresponds to the number of carbon atoms, followed by the suffix –ose.

- Pentoses and hexoses are the most common monosaccharides found in living organisms.

- Disaccharides are carbohydrates that can be decomposed into two monosaccharides. They are composed of two monosaccharides linked together by a glycosidic linkage.

- Polysaccharides are long, chain-like molecules composed of many monosaccharide units.

- Monosaccharides and disaccharides are simple carbohydrates. Polysaccharides are complex carbohydrates.

- Starch, cellulose, and glycogen are polysaccharides. Starch and cellulose are composed of repeating glucose units.

Biochemistry

- Lipids are chemical components of the cell that are insoluble in water but soluble in nonpolar solvents. Lipids are the structural components of cell membranes. They are also used for long-term energy storage and for insulation.

- Fatty acids are lipids with the general formula RCOOH, where R represents a hydrocarbon chain of 3 to 19 carbon atoms.

- Fats and oils are triglycerides. They are triesters, which form from the reaction of glycerol with three fatty acids.

- Phospholipids have the same basic structure as triglycerides, except that one of the fatty acid groups is replaced with a polar phosphate group. The polar part of the molecule is hydrophilic; the nonpolar part of the molecule is hydrophobic.

- Steroids are lipids that have a four-ring structure. Cholesterol, testosterone, and estrogen are common steroids.

- Amino acids are molecules containing an amine group, a carboxylic acid group, and a side chain. Amino acids differ from each other only in their alkyl side chains.

- Alanine, serine, aspartic acid, and lysine are amino acids.

- Dipeptides are molecules composed of two amino acids linked together by a peptide bond.

- Short chains of amino acids are called polypeptides.

- Proteins are polymers of amino acids. Proteins usually contain hundreds or thousands of amino acids joined by peptide bonds.

- The primary structure of a protein is the sequence of amino acids in its chain.

- The secondary structure of a protein refers to short-range periodic or repeating patterns often found along protein chains.

- The alpha (α)-helix is a secondary protein structure in which the amino acid chain is wrapped into a tight coil in which the side chains extend outward from the coil.

- The beta (β)-pleated sheet is a secondary protein structure in which the amino acid side chain is extended and forms a zigzag pattern. The peptide backbones of neighboring chains interact with one another through hydrogen bonding to form zigzagged sheets.

- The tertiary structure of a protein consists of large-scale bends and folds due to interactions between the side chains of amino acids that are separated by large distances in the linear sequence of the protein chain.

Chapter 19

- The quaternary structure of a protein refers to the arrangement of chains in proteins. Quaternary structure is maintained by interactions between amino acids on the individual chains.

- Nucleic acids are polymers that contain a chemical code that specifies the correct amino acid sequences for proteins.

- Deoxyribonucleic acid (DNA) and ribonucleic acid (RNA) are nucleic acids. DNA exists primarily in the nucleus of the cell, RNA exists through the entire interior of the cell.

- Nucleotides are the individual units comprising nucleic acids. Each nucleotide has a phosphate group, a sugar, and a base. Nucleotides link together through phosphate linkages to form nucleic acids.

- The order of bases in a nucleic acid chain specifies the order of amino acids in a protein. A sequence of three bases, called a codon, is needed to code for one amino acid. The same codons specify the same amino acids in nearly all organisms.

- Chromosomes are structures within the nucleus of the cell that contain genes. Humans have 46 chromosomes.

- A gene is a sequence of codons within a DNA molecule that codes for a single protein.

- When a cell divides, each daughter cell receives a complete copy of the DNA within the cell's nucleus.

- When a cell synthesizes a protein, the base sequence of the gene that codes for that protein is transmitted to messenger RNA. The messenger RNA moves out to a ribosome, where the amino acids are linked in the correct sequence to synthesize the protein.

Biochemistry

SKILLBUILDER PROBLEMS AND SOLUTIONS

SKILLBUILDER 19.1 Identifying Carbohydrates

Determine which of the following molecules are carbohydrates and classify each carbohydrate as a monosaccharide, disaccharide, or polysaccharide.

Solution:
You can identify carbohydrates as either an aldehyde or ketone with multiple –OH groups attached or as one or more rings of carbon atoms that include one oxygen atom and also have –OH groups attached to most of the carbon atoms. (a) is not a carbohydrate because it has a carboxylic acid and an amine group and no –OH groups. (b) is a monosaccharide. (c) is a not a carbohydrate because it has carbon-carbon double bonds and a carboxylic acid group and no –OH groups. (d) is a disaccharide.

Chapter 19

| SKILLBUILDER 19.2 | **Identifying Triglycerides** |

Identify the triglycerides among the following molecules and classify each as a saturated or unsaturated fat.

Solution:
Triglycerides are easily identified by the three-carbon backbone with long fatty acid tails. Both (b) and (d) are triglycerides. (b) is an unsaturated fat because it contains double bonds in its carbon chains. (d) is a saturated fat because it does not have any double bonds in its carbon chains.

Biochemistry

SKILLBUILDER 19.3 **Peptide Bonds**

Show the reaction by which valine and leucine form a peptide bond.

$$H_2N-\underset{\underset{CH_3}{\overset{H}{\underset{|}{C}}}\!-\!H}{\overset{H}{\underset{|}{C}}}-\overset{O}{\overset{\|}{C}}-OH \quad\quad H_2N-\underset{\underset{\underset{CH_3}{|}}{\underset{H_3C-\overset{|}{C}-H}{\underset{|}{CH_2}}}}{\overset{H}{\underset{|}{C}}}-\overset{O}{\overset{\|}{C}}-OH$$

Valine Leucine

Solution:
Peptide bonds are formed when the carboxylic end of one amino acid reacts with the amine end of a second amino acid to form a dipeptide and water.

$$H_2N-\underset{\underset{CH_3}{\underset{|}{H_3C-\overset{|}{C}-H}}}{\overset{H}{\underset{|}{C}}}-\overset{O}{\overset{\|}{C}}-OH \;+\; H_2N-\underset{\underset{\underset{CH_3}{|}}{\underset{H_3C-\overset{|}{C}-H}{\underset{|}{CH_2}}}}{\overset{H}{\underset{|}{C}}}-\overset{O}{\overset{\|}{C}}-OH \;\longrightarrow\; H_2N-\underset{\underset{CH_3}{\underset{|}{H_3C-\overset{|}{C}-H}}}{\overset{H}{\underset{|}{C}}}-\overset{O}{\overset{\|}{C}}-NH-\underset{\underset{\underset{CH_3}{|}}{\underset{H_3C-\overset{|}{C}-H}{\underset{|}{CH_2}}}}{\overset{H}{\underset{|}{C}}}-\overset{O}{\overset{\|}{C}}-OH$$

$$+ H_2O$$

SKILLBUILDER 19.4 **Complementary DNA Strand**

Show the sequence of the complementary strand for the following DNA strand.

C C A T T G G

Solution:
Draw the complementary strand, remembering that adenine (A) pairs with thymine (T) and cytocine (C) pairs with guanine (G).

Chapter 19

SELF-TEST QUESTIONS

A. Match the following terms with the phrases below.

alpha (α)-helix
amino acid
beta (β)-pleated sheet
cell membrane
cellulose
codon
complementary base
complex carbohydrate
disaccharide
DNA
ester linkage
fatty acid
glycolipid
glycoside linkage
human genome
lipid bilayer

messenger RNA (m-RNA)
nucleic acid
nucleus (of a cell)
peptide bond
phospholipid
primary protein structure
quaternary protein structure
random coil
RNA
saturated fat
secondary protein structure
simple carbohydrate
starch
tertiary protein structure
unsaturated fat

1. Map of all of the genetic material of a human being
2. Part of the cell that contains genetic material
3. Part of the cell that binds its perimeter
4. Link between two monosaccharides
5. Carbohydrate that can be decomposed into two simpler carbohydrates
6. Class of carbohydrates consisting of monosaccharides and disaccharides
7. Carbohydrate composed of many repeating saccharide units
8. Polysaccharide composed of repeating glucose units in which the oxygen atoms joining neighboring glucose units point down relative to the planes of the rings
9. Polysaccharide composed of repeating glucose units in which the oxygen atoms joining neighboring glucose units are roughly parallel with the planes of the rings but pointing slightly upward
10. Lipid consisting of a carboxylic acid with a long hydrocarbon tail
11. Bond that joins glycerol to the fatty acid of a triester
12. Triglyceride in which the fatty acids are saturated
13. Tryglyceride in which the fatty acids are unsaturated

14. Lipid with the same basic structure as a triglyceride, except that one of the fatty acid groups is replaced with a phosphate group
15. Structure formed by lipids in a cell membrane
16. Biological molecule composed of a nonpolar fatty acid and a hydrocarbon chain and a polar section composed of a sugar molecule such as glucose
17. Molecule containing an amine group, a carboxylic acid group, and a side chain
18. Bond formed by the reaction of the amine end of one amino acid with the carboxylic acid end of another amino acid
19. Sequence of amino acids in a protein chain
20. Short-range periodic or repeating patterns often found along protein chains
21. Secondary structure of a protein where the amino acid chain is wrapped into a tight coil in which the side chains extend outward from the coil
22. Secondary protein structure in which the amino acid chain is extended and forms a zigzag pattern
23. Secondary protein structure with irregular patterns
24. Large-scale bends and folds of a protein chain
25. Arrangement of amino acid chains in proteins
26. Acid that contains a chemical code that specifies the correct amino acid sequence for proteins
27. Nucleic acid that exists primarily in the nucleus of the cell
28. Nucleic acid that exists through the entire interior of the cell
29. Sequence of three nucleotides and their associated bases
30. Base that is capable of precise pairing with only one other base
31. RNA that moves out of the cell's nucleus and moves through a ribosome to synthesize a protein

B. True/False

1. Carbohydrates provide short-term energy storage in living organisms.
2. Glucose is a pentose.
3. Starch is a polysaccharide.
4. Fatty acids are carboxylic acids with long hydrocarbon tails.
5. Saturated fats tend to be liquids at room temperature.
6. Proteins are polymers of fatty acids.
7. Glycine is an amino acid.
8. The alpha helix is an example of primary protein structure.
9. Nucleotides consist of a sugar group, a phosphate group, and a base.
10. DNA has a double-stranded helical structure.

C. Multiple Choice

1. Glucose is a
   a) Pentose
   b) Hexose
   c) Heptose
   d) Octose

Chapter 19

2. Which of the following is a polysaccharide?
   a) Fructose
   b) Galactose
   c) Sucrose
   d) Cellulose

3. Vegetable oils are
   a) Lipids
   b) Amino acids
   c) Carbohydrates
   d) Proteins

4. Proteins are polymers of
   a) Monosaccharides
   b) Disaccharides
   c) Fatty acids
   d) Amino acids

5. The bond formed by the linkage of two amino acids is called a
   a) Hydrogen bond
   b) Unsaturated bond
   c) Peptide bond
   d) Complex bond

6. The amino acid sequence in a protein corresponds to the protein's
   a) Primary structure
   b) Secondary structure
   c) Tertiary structure
   d) Quaternary structure

7. In the DNA double helix, cytocine hydrogen bonds with
   a) Thymine
   b) Adenine
   c) Guanine
   d) Histomine

8. The sugar in DNA is
   a) Ribose
   b) Deoxyribose
   c) Glucose
   d) Sucrose

9. Which of the following codes for a single amino acid?
   a) Codon
   b) Chromosome
   c) Gene
   d) Nucleotide

10. The human chromosome contains how many genes?
    a) 26
    b) 46
    c) 106
    d) 1046

D. Crossword Puzzle

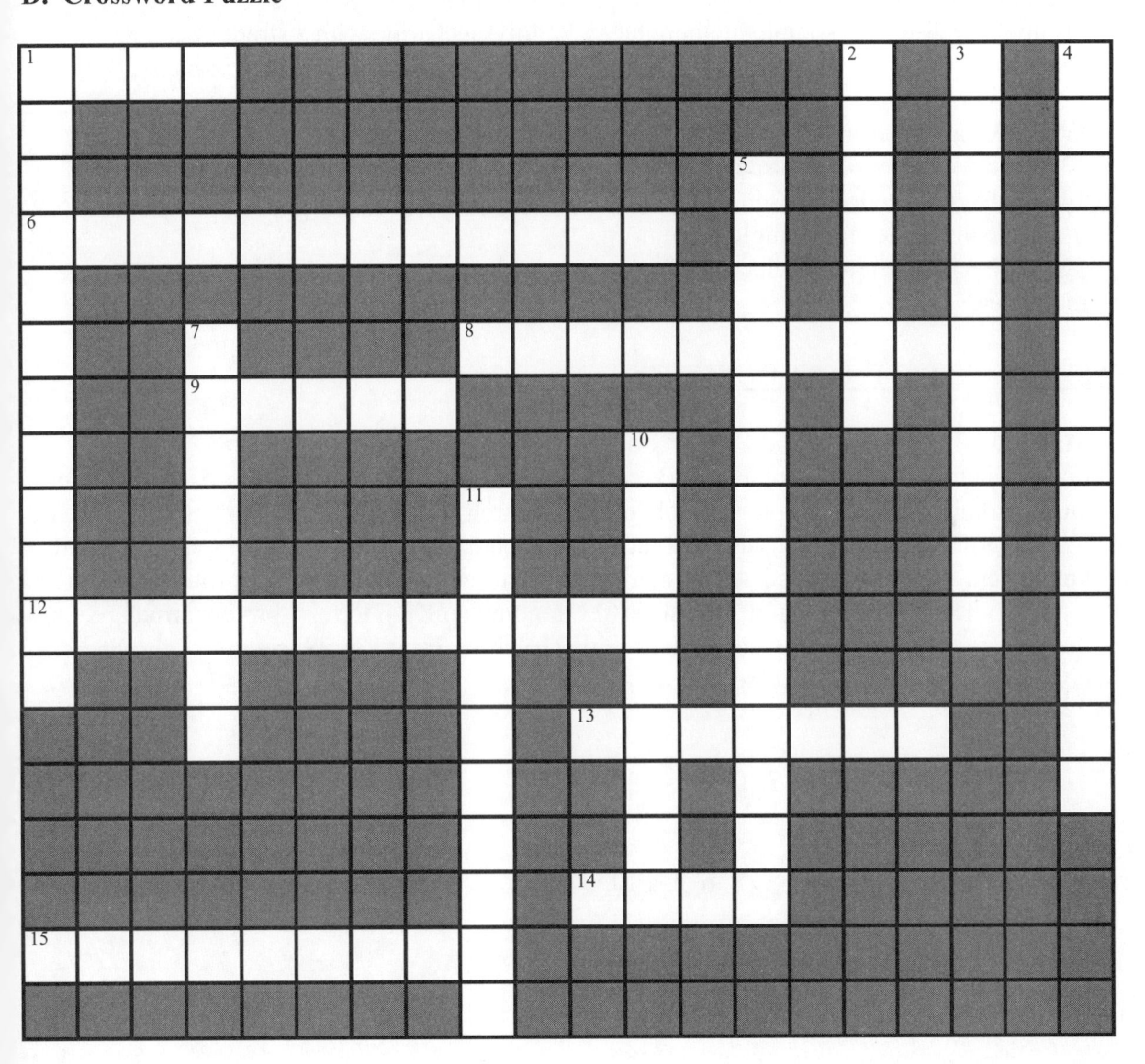

Chapter 19

ACROSS

1. Smallest structural unit of living organisms that has the properties associated with life
6. Study of the chemical substances and processes that occur in plants, animals, and microorganisms
8. Individual unit in a nucleic acid chain composed of a sugar group, a phosphate group, and a base
9. Chemical component of the cell that is insoluble in water but soluble in nonpolar solvents
12. Triester composed of glycerol with three fatty acids attached
13. Lipid with a four-ring structure
14. Sequence of codons within a DNA molecule that codes for a single protein
15. Region between the nucleus and cell membrane

DOWN

1. Primary class of molecules responsible for short-term energy storage in living organisms
2. Polymer composed of amino acids
3. Short chain of amino acids joined by peptide bonds
4. Long, chain-like molecules composed of many adjacent monosaccharide units
5. Carbohydrate that cannot be broken down into simpler carbohydrates
7. Polysaccharide with a structure similar to starch, but the chain is highly branched
10. Molecule that results from two amino acids linking together
11. Structure within the nucleus of the cell that contains DNA

ANSWERS TO SELF-TEST QUESTIONS

A. Matching
1. human genome 2. nucleus (of a cell) 3. cell membrane 4. glycoside linkage 5. disaccharide
6. simple carbohydrate 7. complex carbohydrate 8. starch 9. cellulose 10. fatty acid 11. ester linkage 12. saturated fat 13. unsaturated fat 14. phospholipid 15. lipid bilayer 16. glycolipid
17. amino acid 18. peptide bond 19. primary protein structure 20. secondary protein structure
21. alpha (α)-helix 22. beta (β)-pleated sheet 23. random coil 24. tertiary protein structure
25. quaternary protein structure 26. nucleic acid 27. DNA 28. RNA 29. codon
30. complementary base 31. messenger RNA (m-RNA)

B. True/False
1. T 2. F 3. T 4. T 5. F 6. F 7. T 8. F 9. T 10. T

C. Multiple Choice
1. b 2. d 3. a 4. d 5. c 6. a 7. c 8. b 9. a 10. b

D. Crossword Puzzle

	1	2	3	4	5	6	7	8	9	10	11	12	13	14	15
1	¹C	E	L	L								²P		³P		⁴P
2	A											R		O		O
3	R										⁵M	O		L		L
4	⁶B	I	O	C	H	E	M	I	S	T	R	Y	O	T	Y	Y
5	O											N		E	P	S
6	H		⁷G				⁸N	U	C	L	E	O	T	I	D	E	A
7	Y		⁹L	I	P	I	D				S		N		P	C
8	D		Y						¹⁰D		A			T	C
9	R		C			¹¹C		I		C			I	H
10	A		O			H		P		C			D	A
11	¹²T	R	I	G	L	Y	C	E	R	I	D	E	H		E	R
12	E		E			O		P		A				I
13			N			M		¹³S	T	E	R	O	I	D	D
14						O		I		I				E
15						S		D		D				
16						O	¹⁴G	E	N	E				
17	¹⁵C	Y	T	O	P	L	A	S	M						
18						E									

293